The Impact of Technology on Relationships in Educational Settings

As the linguistic, cognitive and social elements of our lives are transformed by new and emerging technologies, educational settings are also challenged to respond to the issues that arise as a consequence. This book focuses on that challenge, using psychological theory as a lens to highlight the positive uses of new technologies in relationships and educational settings, and to advocate technological learning opportunities and social support where the misuse and abuse of ICT occur.

The Impact of Technology on Relationships in Educational Settings sets out to explore the role of ICTs in relationship forming, social networking and social relationships within our schools, and has grown out of the European Cooperation in Science and Technology (COST) Action on cyberbullying, involving 28 participating countries, and two non-COST countries, of which Australia is one.

This cutting-edge international text offers cross-cultural psychological perspectives on the positive uses of new and emerging technologies to improve social relationships and gives examples of best practice to prevent virtual bullying. This comes at a time when much of the focus in current writings has been on the more negative aspects which have emerged as new technologies have evolved: cyberbullying, cyber-aggression and cybersafety concerns.

This book is ideally suited to researchers and practitioners in the fields of educational and developmental psychology, as well as those specializing in educational technology and the sociology of education.

Angela Costabile is Professor of Developmental and Educational Psychology at the University of Calabria in Italy. Her research centres on network relationships between children of school age and revealed behavioural categories.

Barbara A. Spears is Senior Lecturer in the School of Education at the University of South Australia. A founding member of the Australian Universities' Anti-Bullying Research Alliance (AUARA), her research has examined girls, peer relationships, and covert and cyberbullying practices.

The Impact of Technology on Relationships in Educational Settings

Edited by
Angela Costabile and Barbara A. Spears

Routledge
Taylor & Francis Group
LONDON AND NEW YORK

First published 2012
by Routledge
4 Park Square, Milton Park, Abingdon, Oxon OX14 4RN
605 Third Avenue, New York, NY 10017

Routledge is an imprint of the Taylor & Francis Group, an informa business

British Library Cataloguing in Publication Data
A catalogue record for this book is available from the British Library

Library of Congress Cataloging in Publication Data
The impact of technology on relationships in educational settings / [edited by] Angela Costabile, Barbara A Spears.
p. cm.
1. Education--Effect of technological innovations on. 2. Educational technology--Social aspects. 3. Students--Social networks. I. Costabile, Angela. II. Spears, Barbara A.
LB1028.3.I45 2012
371.33--dc22
2011032616

ISBN: 978-0-415-61007-0 (hbk)
ISBN: 978-0-415-61008-7 (pbk)
ISBN: 978-0-203-13333-0 (ebk)

Typeset in Garamond
by Taylor & Francis Books

Contents

List of illustrations

Figures

Tables

Notes on contributors

Yair Amichai-Hamburger is the Director of the Research Center for Internet Psychology (CIP), located at the Sammy Ofer School of Communications, Interdisciplinary Center, Herzliya, Israel. A prolific author, he has received awards from the Academy of Management and the American Library Association. His research interests include: internet use and well-being; intergroup contact on the internet and virtual teams. His first book, *The Social Net: Human Behavior in Cyberspace*, was published by Oxford University Press in 2005 and an updated edition is due out in 2012. His second book, *Technology and Wellbeing*, was published by Cambridge University Press in 2009.

Bassam Aouil is Associate Professor at Kazimierz Wielki University, Bydgoszcz, Institute of Psychology, Poland, a psychologist and sexologist, and an activist for the promotion of mental health online. He is author and editor of 11 monographs and over 70 articles in the field of clinical psychology, psychology of rehabilitation, pedagogy and sexology.

Haukur Arnthórsson is an independent adviser in Reykjavik, Iceland, in the field of e-government, a part-time teacher at the University of Iceland in e-government and organizational theory, and a scholar at the Icelandic Centre for Research. He holds a PhD from 2008 from the Faculty of Political Sciences at the University of Iceland in the field of public administration, specifically, e-government, its premise and impact. He is the founder of the company Net-Id in Reykjavik which specializes in net address systems.

Maria Giuseppina Bartolo gained a PhD in Child Development from the University of Minho, Portugal, in 2006. She has collaborated on international research projects (TMR, DAPHNE) on peer interaction in primary schools. Her research focuses on bullying and cyberbullying from infant school through to high school. She is currently involved in many research projects at the University of Calabria, Italy, and teaches Techniques of Observation of child behaviour in the course Primary Teaching Education.

Antonella Brighi is a Researcher and Lecturer in Developmental Psychology in the Faculty of Education, Department of Psychology, University of Bologna, Italy. She has collaborated on international research projects on peer interaction and social inclusion in preschool and school years. She is currently coordinating a European Research project on cyberbullying in adolescence; and is a partner in an international research network on the positive and negative uses of technology (COST Action IS0801).

Marilyn A. Campbell is currently an Associate Professor at the Queensland University of Technology, preparing psychologists and school counsellors in the Masters of Education programme. Prior to this, Marilyn supervised school counsellors, and has worked in infants, primary and secondary schools as a teacher, teacher-librarian and school counsellor. Her main clinical and research interests are the prevention and intervention of anxiety disorders in young people and the effects of bullying, especially cyberbullying, in schools. She is the author of the Worrybusters series of books for anxious children.

Angela Costabile is Full Professor of Developmental Psychology at the University of Calabria, Italy. She was delegated by the Rector to the Students' Orientation and is Team Leader on several national and international projects. She is the Working Group Chair and Committee Member of the COST Action IS0801. Her main research fields are: mother–child relationships; and bullying and cyberbullying among students. She is the author and co-author of a large number of books and scientific articles.

Greg Cox is a senior freelance consultant to and former Senior Policy Advisor in Behaviour Management in South Australia (2000–10), providing advice to principals, the Chief Executive and the Minister of Education in the area of behaviour management, including bullying and cyberbullying in schools. A graduate with secondary teaching qualifications from what is now the University of South Australia, Greg has a Graduate Certificate in Gifted Education from the University of New South Wales, and a Master of Education (Educational Administration) from Adelaide University. His teaching career has included roles in Outdoor Education, Special Education (students with behavioural issues), student counselling, Aboriginal Education, and he was an Assistant Principal before assuming the advisory role to the Minister.

Rosario Del Rey is Associate Professor in the Department of Developmental and Educational Psychology in the University of Seville, Spain, and a member of LAECOVI (Laboratory of Studies about Convivencia and Violence Prevention), which is managed by Professor Rosario Ortega. Her work as a researcher has focused on the study, comprehension and intervention of *convivencia* in schools and its difficulties (school violence, bullying and cyberbullying), and she has published widely in this area. She participates regularly in national and international research projects on this issue, such as Cyberbullying (COST IS0801) and is helping public administration (i.e. the Basque Country) to cope with problems and to improve *convivencia*.

Enrico de Santo is at present a lecturer and coordinator of MEDCALABRIA, Centro di Media Education, Calabria, Italy. He gained his PhD from the University of Calabria. His scientific interest is mass media, communicative processes and media education. His most recent writings focus on the virtual power of mobile telephony; media-morphosis and educational issues; media education; media and loneliness; and media and training.

Thaddaeus Douglas is a member of the Policy and Research Team, Beatbullying (UK); webpage: www.cybermentors.org.uk.

Manuela Fabbri is a researcher in Didactic and Special Education in the Faculty of Education, Department of Education, University of Bologna, Italy. She teaches Didactics, Technology of Education and Media Education in Social Education courses. She is involved in national and international research projects mainly in the field of e-learning environments and learning objects, educational innovation and ICT, socio-constructivist learning and social networking, new technologies and inclusive learning environments and empowerment and new technologies for the prevention and treatment of addiction.

Kevin Glasheen has worked in the education sector for thirty years. He has been a teacher, primary school Principal and for the last eighteen years has worked as a guidance counsellor in Australian government secondary schools. His interests focus on help-seeking behaviours and the counselling of young people. He is currently conducting research into the provision of online counselling within the secondary school as part of his PhD studies at the Queensland University of Technology.

Luigi Guerra is Principal Professor of Didactic and Special Education, Dean of the Faculty of Education at the University of Bologna, Italy. He is responsible for several national and international research projects and scientific studies in the field of didactic innovation and ICT; ICT in children's daily life; socio-constructivist learning and educational innovation; and ICT and inclusive learning environments.

Zack Hayat is a fourth-year PhD student in the Faculty of Information at the University of Toronto, and a student collaborator at NetLab, at the University of Toronto. He holds an MA in Mass Communication from the University of Haifa, Israel, where he specialized in the social structure of online communities, specifically on the motivations for participating in such communities. His research examines the role of the internet in maintaining relationships and accessing resources.

Ari Kaukiainen is a lecturer and a student counselling psychologist at the University of Turku, Finland. He gained his PhD from the University of Turku. He has been a special researcher, material developer and a user-instructor of the KiVa antibullying program. He is currently training to be a professional cognitive psychotherapist, specializing in victims of school bullying. His research interests include antecedents of indirect manipulative aggression; connections between learning disabilities and bullying; and bullying among university students. He has co-authored publications in international and national books and refereed journals, such as *Aggressive Behavior*, *Scandinavian Journal of Psychology*, and *Personality and Social Psychology Bulletin*.

Jette Kofoed is Associate Professor in the Department of Education, University of Åarhus, Denmark. She is co-directing the research project eXbus, Exploring Bullying in Schools, and is responsible for the research on cyberbullying. She has published widely on the subjects of processes of inclusion and exclusion among schoolchildren and currently focuses on affectivity in cyberbullying and the agency of technologies in cyberbullying.

Henna Lahti is a Master's student in the Department of Communication at the University of Jyväskylä, Finland. She is currently completing her Master's thesis on technologically-mediated peer communication of children and youth, focusing on how they maintain and form these peer relationships. She also works as a communications officer in Aalto University Student Union. Among other positions, she has served as a board member of the Student Union in the University of Jyväskylä, being responsible for PR, communications, and the student newspaper.

Ersilia Menesini is Professor of Developmental Psychology at the University of Florence, Italy. Her main areas of research are: psycho-social risk factors in the school context (bullying and cyberbullying); romantic relationships, couple aggression and risk behaviour in adolescence; the development of moral emotions and moral sensibility; and the development and evaluation of school interventions against bullying.

Annalaura Nocentini is a post-doctoral researcher at the University of Florence, Italy. Her research interests are mainly focused on aggressive behaviour during adolescence (e.g., dating aggression, cyberbullying, and sexual harassment); and psychosocial adjustment.

Rosario Ortega is Full Professor, Chair of Psychology and Head of the Department of Psychology, University of Cordoba, Spain. She has been a Member of the National Commission for Evaluation of Scientific Projects in Psychology and Education (ANEP) and the National Commission for Evaluation of Professorship and the National System (ANECA); as well as the Permanent Commission of the State Observatory of School Convivencia

(Spanish). She is a Senior Member of the European Association for Research on Learning and Instruction (EARLI) and Team Leader for several national I+D Projects and European and international projects and research Net-working, such as Cyberbullying (COST IS0801). She is the author and co-author of a large number of books and scientific articles about children's play, *convivencia*, bullying, cyberbullying and dating violence among young people.

Elena Pacetti is a researcher in Didactic and Special Education in the Faculty of Education, Department of Education, University of Bologna, Italy. She teaches Didactics and Technology of Education in the Primary Teaching Education course in Bologna. She is involved in several Italian and international research projects in the fields of: didactic innovation and ICT; socio-constructivist learning and social networking; ICT and inclusive learning environments; ICT in children's daily life; and university competences for social professions.

Anna Lisa Palermiti is a researcher in Developmental Psychology at the University of Calabria, Italy, where she teaches the Psychology of Development and Social Psychology. She has collaborated on international research projects on positive and negative relationships between peers. She is involved in many research projects at the University of Calabria, focusing on bullying and cyberbullying from infants' school to high school.

Vera Popper studied psychology at the University of Graz in Austria. Her interests are in evaluation research and organizational development. She is currently working as a teaching and research associate (at PHD student level) in the Faculty of Psychology in the Department of Economic Psychology, Bildung-Psychology and Evaluation at the University of Vienna.

Maili Pörhölä is a Professor in the Department of Communication at the University of Jyväskylä, Finland, where she is the Director of the Mental Violence in Communication Relationships at School and in the Workplace research project. Her research focuses on bullying at school, in higher education, and in the workplace, including examining aspects of peer relationships, peer community integration, and psycho-social well-being. She has published her research widely, as well as authoring several book chapters. In addition, she has co-edited the book *Anti and Pro-social Communication: Theories, Methods, and Applications*, and co-authored the book *Bullying: Contexts, Consequences, and Control*, which has been translated into Spanish and Catalan.

Elisa Poskiparta is the leader of the business unit of the Centre for Learning Research at the University of Turku, Finland. Since 2006, she has been co-leading the development, evaluation, and implementation of the KiVa antibullying program. Her research interests include beginning reading;

risk factors and remediation of reading disabilities; and bullying and victimization in schools.

Virpi Pöyhönen is a doctoral student of psychology at the University of Turku, Finland. In her doctoral thesis she is examining which factors enable some students to defend victimized classmates. In addition to her doctoral studies she has, since 2006, worked on the KiVa antibullying program, developing the materials and educating teachers.

Jacek Pyżalski is Assistant Professor at the Institute of Occupational Medicine in Łódź, and a Lecturer at the Pedagogy Academy in Łódź, Poland. He is a coordinator and researcher on many international and national research programmes devoted to the inappropriate behaviour of students; teachers' occupational stress; and cyberbullying. He is the Polish representative in the COST IS0801 Action dedicated to electronic aggression among children and adolescents and the author of the first Polish monograph on cyberbullying (2010).

Christina Salmivalli is a Professor of Psychology at the University of Turku, Finland. She gained her PhD from the University of Turku. Since 2006, she has been co-leading the development, evaluation, and implementation of the KiVa antibullying program. She is also PI for several research projects funded by the Academy of Finland. Her research interests include children's and adolescents' peer relations, with a special focus on peer-to-peer bullying. She has published extensively in books and in high-ranked refereed journals, such as *Child Development*, *Developmental Psychology*, and *Journal of Clinical Child and Adolescent Psychology*.

Virginia Sánchez is Associate Professor in the Department of Developmental and Educational Psychology in the University of Seville, Spain, and a member of LAECOVI (Laboratory of Studies about Convivencia and Violence Prevention), which is managed by Professor Rosario Ortega. Her research has focused on the study of emotional development of bullies and victims and, in the past few years she has been involved in the study of dating violence and sexual harassment in adolescence. She has published scientific articles related to both topics. At present, she is interested in the study of cyberdating, trying to relate the use of social networks and adolescents' romantic relationships. She is participating in two national and international research projects on Cyberbullying (COST IS0801).

Maria Sapouna is Visiting Research Fellow in the Department of Psychology, University of Warwick, UK. Her expertise lies in the field of developmental criminology. She is experienced in using longitudinal data to investigate the development of delinquency, bullying and victimization over time. Her other main research interest is in the evaluation of interventions to tackle aggression, bullying and violence.

Phillip T. Slee is Professor in Human Development in the School of Education, Flinders University, South Australia, and Director of the Flinders Research Centre for 'Student Wellbeing & Prevention of Violence' (SWAPv). Phillip is a trained teacher and registered psychologist. He has published extensively in the field of child development, produced educational resources and is interested in the practical and policy implications of research. His chief areas of research interest include: child and adolescent mental health, childhood bullying/aggression, family development, conduct disorders, stress and teacher education. Phillip's research team has evaluated the KidsMatter Primary and Early Childhood Mental Health Initiatives: http://caef.flinders.edu.au/kidsmatter/. His website is http://www.caper.com.au.

Barbara A. Spears has taught Educational Psychology in the School of Education at the University of South Australia since 1996. She gained her PhD from Flinders University. Her research has focused on girls' peer relationships, friendships and bullying behaviours and most recently covert bullying, cyberbullying, bystanders and youth voice. She is co-Director of the Citizenship and Wellbeing research group of the Centre for Research in Education; has been team leader for the Australian Government qualitative study into Covert Bullying; and is currently a senior researcher with the Australian Government Collaborative Research Centre for Young People, Technology and Wellbeing (YAW-CRC): http://www.yawcrc.org.au/.

Christiane Spiel is Professor of Educational Psychology and Evaluation and Department Head in the Faculty of Psychology, University of Vienna, Austria. She is and has been chair and member of various international advisory and editorial boards, such as e.g., President of the European Society for Developmental Psychology, and President of the DeGEval, Society for Evaluation. In several projects she has worked with Federal Ministries in Austria. She has received several awards and has published more than 200 original papers. Her research topics are: bullying and victimization; lifelong learning; integration in multicultural school classes; gender stereotypes in education; and evaluation research.

Dagmar Strohmeier studied psychology, philosophy and Turkish language and culture at the University of Graz in Austria. She holds a PhD in psychology from the University of Vienna, Austria. Between 2006 and 2011, she was Assistant Professor at the University of Vienna. Currently, she holds a chair in intercultural competence at the Upper Austria University of Applied Sciences, School of Health/Social Sciences. The main topics in her research are bullying and victimization, homophily in friendship choices, and the development, implementation and evaluation of a school-based programme to prevent bullying and victimization in youth (the ViSC programme).

Jessica von Kaenel-Flatt read English Literature and Theology at Cambridge University. She has travelled widely and lived in five countries around the world and has worked for Beatbullying since January 2009; webpage www.cybermentors.org.uk.

Dieter Wolke is Professor of Developmental Psychology in the Department of Psychology, University of Warwick, and Professor in the Department of Psychology and Health Sciences Research Institute at Warwick Medical School, UK. His teaching is on Developmental Psychopathology. He has worked at the Universities of Bristol, Hertfordshire, Munich and London and in the research funding sector (Jacobs Foundation, Zurich). His research is interdisciplinary and longitudinal with a focus on biological (e.g. preterm birth) or environmental risks (e.g. bullying) and the development of psychopathology. He is involved in several large-scale studies including the Bavarian Longitudinal Study and Understanding Society. He is on the editorial boards of several paediatric or psychology journals.

Alison J. Wotherspoon graduated with a BA DipEd from the University of New South Wales in 1982, began her career in television at the BBC, worked at Film Australia, the ABC, and SBS and studied Producing at AFTRS, the Australian National Film School. She worked as an independent producer before moving to Adelaide in 1997 to teach Screen Production at Flinders University. Alison has produced and directed a number of series about bullying and young people and her work *Come into My World*, about Alzheimer's disease, was nominated for an ATOM award in 2009. Alison is a corresponding member of CILECT and an Executive Member of ASPERA, and CAPA (CILECT Asia Pacific Association); http://www.flinders.edu.au/people/alison.wotherspoon.

Foreword

I am very pleased to write a short foreword to this book, *The Impact of Technology on Relationships in Educational Settings*. It is one of the outcomes from a COST Action Network on Cyberbullying, which started in October 2008 and is continuing for four years. COST stands for Co-operation in the field of Scientific and Technical Research, and a COST Action funds meetings, workshops and seminars, scientific visits, training schools, and dissemination in its particular area.

This particular COST Action, IS0801, is entitled: *Cyberbullying: Coping with Negative and Enhancing Positive Uses of New Technologies, in Relationships in Educational Settings*. At the time of writing, there are 28 European COST countries involved in the Action, plus Australia and Ukraine as non-COST countries. The Action website is at: http://sites.google.com/site/costis0801/.

One of the editors of this book, Professor Costabile, is from Italy. Italian researchers have contributed strongly to the Action, leading two of our six Working Groups, and Angela Costabile has led one of these, focusing on positive uses of new technologies in relationships in educational settings. This book is the major output from this Working Group. The second editor, Dr Spears, is from Australia. Australian colleagues too have made a huge contribution, including procuring joint funding of a Training School for early career researchers, held in Melbourne in 2010.

This book brings together contributors from some ten different countries, most of whom (not all) are members of the COST Action. After Part I sets the background for use of ICT by young people and in schools, the book develops two main aspects. One is the positive use of ICT generally, in educational settings such as schools, for example, in promoting relationships, or providing counselling. The other focuses more on cyberbullying, the main negative aspect of ICT use. However, Part I looks at ways of reducing, or ameliorating the effects of, cyberbullying, again using ICT to do this.

There are now quite a number of books that focus on cyberbullying, and on the relevant research in this area. However, this book breaks rather new ground in covering the positive and prosocial uses of ICT in schools, while also being aware of the negative aspects, and using ICT to tackle these. These are problems with an international dimension, just as mobile phones and the internet are worldwide phenomena. The range of this book, both cross-nationally and in terms of the topics covered, will ensure that it is a very important contribution to the research community, but also of immediate practical relevance to educators, parents, and others concerned with young people's well-being in educational settings.

Peter K. Smith
Chair, COST IS0801
London
March 2011

Acknowledgements

We are grateful to the members of COST Action IS0801 *Cyberbullying: Coping with Negative and Enhancing Positive Uses of New Technologies, in Relationships in Educational Settings* for providing the impetus and support to compile this volume.

We wish to thank Maria Giuseppina Bartolo for her cooperation and assistance in the preparation and finalization of this book.

Introduction

Barbara A. Spears and Angela Costabile

New and emerging information and communication technologies (ICTs) increasingly are presenting society with significant opportunities and challenges. The ease and convenience of the communication is counterbalanced by concern regarding the safety and well-being of our young people as they operate in this ICT-driven environment. As young people rush to embrace ICTs and social media in particular, engaging with each other across boundary-less environments, educational settings not only are engaging in strategies to support the use of technology in contemporary teaching and learning approaches, but are also seeking ways to support young people in their relationship growth and development in a technological era.

Social media are re-writing how we interrelate socially and are forging reconsiderations and reassessments of social boundaries and the relationships which operate within and around them. As the lines between offline and online behaviours are becoming increasingly blurred, educational settings are challenged not only to respond in a reactive sense to the safety issues that arise through the misuse and abuse of new technologies: through specific interventions or post-ventions, but also to take proactive action to ensure that the positive uses of technologies are acknowledged and championed, and are viewed as enablers of young people and their well-being.

This book is a product of a wider collaboration between many different countries (28 European and 2 non-COST countries, Australia and Ukraine) involved in a European Co-operation of Science and Technology (COST) Action IS0801 entitled *Cyberbullying: Coping with Negative and Enhancing Positive Uses of New Technologies, in Relationships in Educational Settings.*

New media and technology are the latest vehicles for the complex range of human behaviours and interactions we have come to recognize as bullying. It must be remembered, that bullying is always concerned with a *social relationship*, where there is an imbalance of power, repetition and a desire to harm or hurt. *Cyberbullying*, the term used to describe bullying which occurs through the use of technology, has become a global phenomenon in a very short period of time, and impacts on individuals, particularly young people, in a range of negative ways, including, in some cases, leading to suicide. These impacts are not unique to any one country.

While technology constantly and rapidly evolves around us, the behaviours associated with bullying change little, rather, they have adapted to these environments, crossing the boundaries of time and place through the 'always on' environment and cycling between home and school and back again, creating an air of inescapability for those being targeted.

The seeming rising moral panic among some parents, educators and policy-makers which accompanies the notions of cyberbullying and cybersafety, often calls for technology to be altered, filtered, removed, or restricted. There is a real need therefore, to overcome the initial knee-jerk reaction that positions technology as the problem when thinking about cyberbullying and cybersafety issues.

Rather, this book positions technology as part of the solution: as a means to an end; a way to support human behaviours positively and to provide avenues for intervention. This book serves as a counterbalance: to ameliorate and alleviate that panic by outlining an evidence base that demonstrates the *positive* ways that technology is being and can be used to impact on relationships in educational settings.

Regardless of how relationships are enacted as young people grow and develop, or how bullying morphs over time, technology is here to stay and is impacting on society and the ways in which we interact in ways we had never previously considered. The rapid progression from Web 1.0 to Web 2.0 platforms: from email and SMS to social networking sites, to Twitter and virtual worlds; from being passive recipients to creators and publishers of content online; from having stand-alone PCs to small, hand-held internet-ready devices already indicates that there is more yet to come. New complex social issues arising from the ways ICTS are being used, require new, equally complex social solutions. This book provides an international lens through which to view the impact of technology on the ways young people interact now, and sets the scene for future research.

Leading educational, developmental and technological experts from many countries have contributed chapters in order to capture how technology is being used in positive ways in their cultural context, to deal with relationships and cyberbullying across educational settings. The different approaches and disciplines contribute two different perspectives: theoretical and interventionist, making it of interest to communities, practitioners and policy-makers. The text brings diverse examples of practices already being employed in different countries, to a wider, global audience.

Cyberbullying is a global issue, and will require the collective knowledges from across the globe to deal with it. This book is the first step beyond simply describing the phenomenon. Bullying has come to the global village through the advent of technology and will require technology to be part of the solution. Relationships, however, exist across and through technology.

This book focuses on that challenge: to highlight the positive uses of technologies in relationships in educational settings, which will inform practice and shape policy, and to advocate for social supports and learning opportunities which can be delivered *through* technology.

The main objective of this book is to present a counterpoint to those books that already deal with understanding and describing the phenomenon of cyberbullying. This text offers a cross-cultural, international perspective on the positive uses of new and emerging technologies to improve social relationships and provides examples of current evidenced-based practices to prevent and intervene with virtual bullying. Our goal is to facilitate those involved in educational settings (such as parents, teachers, educators) to recognize that ICTS in relationships in educational settings can enable and empower young people and their wellbeing.

This book is comprised of 17 chapters, with contributions from leading researchers from Europe, Israel, and Australia. While each nation has its own unique cultural identity and education system, new technologies cross all boundaries of time and place, making this book an exciting cross-cultural evidence base of the positive uses of new and emerging technologies in relationships in educational settings, at a time when much of the focus in current writings has been on the more negative aspects which have emerged as new technologies have evolved: cyberbullying; cyberaggression and cybersafety.

Part I, 'Positive uses of ICTs: background issues' (Chapters 1–3) presents some of the main background issues which together form a foundation from which to examine how new and emerging technologies are being used to support and enable individuals in their relationships. In order that the book is grounded in current evidence, Barbara A. Spears, Jette Kofoed, Maria Giuseppina Bartolo, Anna Lisa Palermiti and Angela Costabile, in Chapter 1, present young people's perspectives of the positive uses of social networking, from three different countries. Haukur Arnthórsson in Chapter 2, challenges readers to consider secure communications on the internet and the role that a system of national net addresses could play in protecting identity, and Enrico De Santo and Angela Costabile in Chapter 3, outline a history of media education and highlight the challenges of creating new academic curricula, where a media education can impact positively on the development of good practices in educational contexts.

Part II, 'Scholarly settings' (Chapters 4–11) examines the positive uses of ICTs in scholarly settings. Here the role they play in enhancing relationships in educational settings is highlighted, with a view to promoting the positive impact that technology can have on the lives of children and young people. Antonella Brighi, Manuela Fabbri, Luigi Guerra and Elena Pacetti in Chapter 4, present a discussion of the potential for ICTS to transform school structures and the educational experience, and in Chapter 5, Maili Pörhölä and Henna Lahti report on the main findings of an extensive literature review of the ways in which children and adolescents use ICTs in establishing and maintaining peer relationships, and specifically report on two 'virtual rooms' created for children and adolescents.

Following these initial chapters, three countries report on internet use: for Spain, Rosario Del Rey, Virginia Sánchez and Rosario Ortega, in Chapter 6, report on adolescents' prosociality on the internet and initiatives to deal with bullying; for Austria, Vera Popper, Dagmar Strohmeier and Christiane Spiel in Chapter 7, argue for the use of notebooks and e-learning in schools as a positive example of ICTs' use in school contexts; and for Poland, Jacek Pyżalski, in Chapter 8, examines the digital generation divide between digital 'natives' and 'immigrants' and the functional and dysfunctional styles of internet use in adolescents.

The final three chapters of Part II (Chapter 9–11) explore the psychological impacts and the role that the internet may play in either contributing or supporting individuals. In Chapter 9, Zack Hayat and Yair Amichai-Hamburger report on a literature review of the relationship between internet use and the psychological and psychosocial well-being of children; in Chapter 10, Bassam Aouil reflects on the impact of online support/help in psychological and pedagogical practices, and, in Chapter 11, Kevin Glasheen and Marilyn A. Campbell examine the ways that online school counselling can enhance relationships and connectedness in schools.

Part III, 'Pervention of cyberbullying', brings together some of the international practices and examples of evidence-based practices in using online techniques and strategies to prevent, and intervene in, cyberbullying, as well as those technological practices which encourage participation, support learning opportunities and provide avenues for young people to have agency and voice. In Chapter 12, Ersilia Menesini and Annalaura Nocentini examine the relationship between online and face-to-face peer-led education and intervention strategies; Chapter 13, by Jessica von Kaenel-Flatt and Thaddeus Douglas, reports on a Cybermentoring project, an innovative model of peer mentoring that takes place both online and offline; Chapter 14, by Elisa Poskiparta, Ari Kaukiainen, Virpi Pöyhonen, and Christina Salmivalli, presents young people's experiences of an anti-bullying computer game from the KiVa programme; Chapter 15, by Dieter Wolke and Maria Sapouna, reports on a virtual intervention game to reduce bullying and victimization; Chapter 16, by Alison Wotherspoon, Greg Cox and Phillip Slee, considers the role of using mobile phone technology to counter cyberbullying through sending anti-bullying video and sms messages to other students, and finally, Chapter 17, by Barbara A. Spears, reviews online programs employed in Australia to promote cybersafety and digital citizenship and to address cyberbullying.

Finally, the Conclusion summarizes the key themes presented by this diverse group of authors and presents the reader with challenges to consider and directions for the future.

Part I

Positive uses of ICTs

Background issues

Chapter 1

Positive uses of social networking sites

Youth voice perspectives

Barbara A. Spears, Jette Kofoed, Maria Giuseppina Bartolo, Anna Lisa Palermiti and Angela Costabile

Social networking sites (SNSs) present many challenges and opportunities for young people. These include: managing personal information; associated issues of identity formation and change; grappling with concerns of cybersafety and cyberbullying, and recognizing the legal and copyright implications of their behaviour online. This chapter explores young people's perspectives of the positive uses of technology, in particular, SNSs, to support their relationships, and how those friendship-driven practices (Ito *et al.*, 2008: 10) are played out online on such SNSs as Facebook. A qualitative, cross-national pilot study was undertaken in Australia, Denmark and Italy, to enable young people to share their understandings and experiences, contributing youth voice to this field of enquiry.

In accessing youth voice, this chapter pursues the *meanings* that youth attach to the use of SNSs as they conduct their relationships in and around educational settings. This chapter adopts an interpretative analysis whereby the individual and collective lived experiences are examined, in order to further understand how young people are making sense of their personal and social cyber-configured world (Smith and Eatough, 2007; Storey, 2007).

Background

In recent years, electronic and virtual environments have become increasingly important in the social lives of adolescents (Williams and Guerra, 2007). In the relatively few years since the internet has been mainstream, it has become a major tool of socialization and recreation for both children and adolescents (Rivoltella, 2006). Technologies make it possible to collect and exchange information with peers; to interact with others while maintaining anonymity; to experiment with new identities; to develop a sense of community and to explore social acceptance. It is possible to share opinions through blogs, or socialize and network through social networking sites such as Facebook. Electronic devices have become important players in socialization, and communication skill development for young people involves learning how to navigate interpersonal relations online as well as offline, and back again, in an almost seamless transfer. This so-called timeless and boundary-less place, however, re-introduces time and space in new ways, with its synchronous and analogous abilities co-existing. SNSs are characterized by online activities such as: the establishment of personal profiles; the visibility of social relations; and the option of meeting, maintaining, changing or dismantling social relations. boyd and Ellison (2007, cited in Larsen 2010, p. 10) further define social networking sites as:

> web-based services that allow individuals to: 1) construct a public or semi-public profile within a bounded system; 2) articulate a list of other users with whom they share a connection; 3) view and traverse their list of connections and those made by others, within the system.

According to boyd (2007, p. 7), social network sites such as Facebook operate as '*networked publics*: the spaces and audiences which are bound together through technological networks' and these provide opportunities for youth to explore and make sense of their own identities, status and cultures in similar ways to those public spaces which are offline, such as parks and malls. While supporting many of the same practices, the structural differences between online and offline 'publics' often inflect practices in unique ways. This point is shared in cyberbullying research (Spears *et al.*, 2008; 2009; Kofoed, 2009, Kofoed and Ringrose, forthcoming), where youth learn how to navigate in and between networked and unmediated public domains, such as school settings and SNSs.

The basic premise of *Web 2.0* (post 2004), however, as distinct from *Web 1.0*, is that it allows *users* to create content and share objects within a virtual community, giving rise to the online social interaction and dialogues which young people now take for granted. This aspect serves to separate young people from the previous generations of online users, and places them in a unique place in terms of relationship maintenance and management. The relatively recent development of online tools and platforms, such as blogs, wikis, podcasts and social network sites, offers new ways of communicating and this technologically-mediated communication obviously allows both positive and negative use: to support relationships, or to denigrate them. While much is known about the negative aspects of socializing online, such as cyberbullying, it is equally important to comprehend youth perspectives of the positive ways that SNSs and technology in general support the growth and development of peer relationships and promote positive interactions and skills.

Youth use of social networking sites

According to the Pew Internet Project Report (Lenhart *et al.*, 2010), in September 2009, 73 per cent of American teens aged 12–17 who were online, were using a social networking site. By comparison, the Click and

Connect report of young Australians' use of online social media (Australian Communications and Media Authority, 2009) found that 90 per cent of 12–17-year-olds and 97 per cent of 16–17-year-olds use SNSs, and further reported that these young people were mostly chatting to friends known to them in the real world. Danish children, on average, start using the internet at the age of 8, and 98 per cent of all children have access to the internet in their homes, with 75 per cent having access in their own rooms. Furthermore, 75 per cent of Danish children have a profile on a SNS (EU Kids Online II: www.eukidsonline.net). In Italy, Save the Children in (2008) found that 74 per cent of teen (11–14 years old) children use SNSs to stay in contact with their regular friends. While it is evident that many young people are socializing online, the recent cross-European Study of children's internet use (EU Kids Online II) found that there were a number of national differences, making it impossible to draw a completely clear picture of European children's use of online media. Obviously, as outlined by EU Kids Online II (Livingstone *et al.*, 2011), cultural contexts are very important, and this chapter specifically explores young people's perspectives from three culturally different contexts, seeking commonalities and differences in the *meanings* of their experiences in terms of positive impact. What remains constant for each context, however, is the peer network which, in terms of development and sociality, is important (Haavind, 2000; Staunæs, 2004).

Social peer relations are paramount in the everyday lives of contemporary youth everywhere, and constitute a defined network of contacts that remains fluid, transient and dynamic as relationships ebb and flow across online and offline environments. Social networking sites, such as Facebook, are recently developed, mediated environments, where young people communicate with existing relationships, form new ones, cement ties with others, and re-establish old friendships. How these SNSs contribute to and support positive, friendship-driven practices, is the focus of this chapter.

Youth voice

The United Nations Convention on the Rights of the Child (1989) (http://www.unicef.org/crc/index_30177.html) highlights children's right to participate, stating that they should be free to express their views and to be heard (Article 12) and that they have the right of freedom of expression, and freedom to seek and impart information through any media of the child's choice (Article 13). West (2004) suggests that 'voice', however, is not simply concerned with the opportunity to communicate ideas and opinions, but that it is about having the power to influence change. Traditionally, the views of children and young people were accorded little legitimacy or value in comparison to those of adults, however, insofar as contemporary society is concerned, and their use of SNSs and technology in particular, their views and opinions are of great importance, for SNSs can give us insight into how children's online lives are experienced and what meaning they give to technology. In supporting young people giving voice to their experiences and understandings about technology, and in particular, those aspects of current technology which support positive outcomes for their relationships, we can begin to co-construct meaning around how SNSs are being employed by them.

If, as Prensky (2009) suggests, the 'digital natives are now deemed to have digital *wisdom*', a wisdom which arises *from* the use of technology, then accessing youth voice in terms of that wisdom is fundamental to understanding how young people use technology in positive ways, for their relationships, in educational settings and the meanings they give to it.

Methodology for accessing youth voice

Introduction and sampling

Interpretative analysis aims to explore the lived realities of participants and their experiences of particular phenomena: in this case, how positive uses and impacts of SNSs affect relationships in educational settings. In order that consistency be attained across the three countries (Australia, Denmark and Italy), a methodology plan was established which did not require instruments to be translated, but outlined a qualitative approach which could be undertaken in each country's own language; thus, the following strategies were employed: (1) focus groups; (2) visual elicitation material to stimulate conversation; and (3) key guiding questions which the researcher/facilitator could use as a springboard for discussion with the young people.

A small convenience and purposive sample was sought in each country so that the experiences, perceptions and beliefs of young people who use social networking sites could provide rich, thick data. This approach reflects both intensity sampling, which deliberately seeks excellent or 'rich' examples of the issue being investigated (young people who use social networking sites) and maximum variation sampling, which uses the diversity in the sample as a strength (boys and girls from six school settings in three different countries) to ensure that the core themes which emerge, and are common to all, represent particularly valuable and salient, patterns of central importance (Patton, 1990, p. 169).

Participants

Young people aged 14–16 years old from one or more schools from each country were invited to participate in focus groups, of between 6–8 individuals. Details are provided in Table 1.1. All participation was voluntary, with informed consent from parents, school leaders, and the young people themselves, and was in accordance with the relevant ethical processes and approval of the individual country.

Table 1.1 Participants and methodology of the sample

	Participants	*Method*	*Schools*
Australia	15 boys; 15 girls (N = 30) Aged 14–16 years Grades 9–11	Three mixed focus groups	Three schools in South Australia: one independent one government one Catholic
	30 girls; 30 boys (N = 60) Grades 9–11	Four mixed classes Written responses to three key questions	one government school (as above)
Italy	12 boys; 12 girls (N = 24) Aged 14–15 years	Three mixed focus groups	Two public schools in Southern Italy
Denmark	2 boys, 5 girls (N = 7) Aged 14–15	One mixed focus group	One urban public school
TOTAL	N = 121 students	Seven focus groups Written responses	Six school settings in three countries

Materials: visual elicitation sheets, audio recorder, guiding questions

Photo/visual elicitation sheets which had various social networking site logos relevant to the age group and cultural context (e.g. Facebook; MySpace; Twitter; Bebo; Formspring, Arto, etc.) were used as trigger points and conversation starters for the focus groups. Photo elicitation is a technique whereby visual images are employed to stimulate responses (Wagner, 1978, cited in Harper, 2002, p. 15). This form of data collection, widely used in visual sociology, allows the participant to assume the role of an interpreter, shifting the control from the researcher to the participant (http://labsome.rmit.edu.au/liki/index.php/Visual_Research_Method; see also Visual Sociology Association: http://www.visualsociology.org/). Given that social networking sites are inherently visual media, using this form of stimulus was congruent with the topic under investigation. An audio recorder was used to capture the complete discussion.

Procedure

The focus groups were conducted in quiet rooms at the relevant school sites, over the course of approximately one hour. The participants were seated so that they could see each other, and were reminded that the session would be audio-recorded, and that they had the freedom to choose not to answer any questions or not participate at any time, without repercussions. The facilitator explained the purpose of the session, including respecting confidentiality and anonymity. Following a short introduction by each person, participation norms were established by the group, e.g. respecting the views of others. The SNSs' logo sheets were distributed and students were encouraged to talk to a partner initially, before feeding back to the group as a whole, their familiarity, knowledge and participation of the SNSs depicted. The focus group was steered towards the issues under consideration using both narrative and reflexive questions which included descriptions of their everyday lives, which allowed similarities and differences among the participants to emerge. Guiding questions and prompts were used to delve further or to keep the conversations moving. The core guiding questions were:

1 In what ways are social networking sites a part of your daily life?
2 How do you think that using technology and social networking sites is a positive influence in your lives?

In addition, the following questions were also touched upon:

3 How can we use technology to challenge cyberbullying?
4 What do you want adults to know about the ways you use technology and its importance in your lives?

Focus group discussions were audio-recorded so that the researchers could listen, facilitate the group, and make observations freely of non-verbal communications, the group dynamic and the flow of verbal exchange.

Data analysis

Focus group audio-recordings from each country were transcribed and processed by the relevant researcher using spreadsheets and mind maps, where issues and reflections were systematically examined for core themes: existing knowledge (*a priori*) and those which emerged naturally from the groups in response to the guiding questions and from the dis/agreements among the participants (inductive). The general principle followed was one of data reduction and theme generation (see Miles and Huberman, 1994; Kofoed, 2004). The coding processes were done in the native languages. In the Danish and the Italian cases, quotes and core themes were translated into English for shared analytical purposes. The analytical process involved several approaches: *open coding*, whereby words and phrases were highlighted and colour coded, to determine: (1) known/existing themes (*a priori* coding); or (2) new themes which emerged from the transcript. A process of abstraction and deduction, and continuous comparison with existing knowledge and within and across the groups followed (*axial coding*), which identified relationships and grouping of themes and ideas. *Selective coding* was then used to draw illustrative examples of the central core experiences and understandings of the participants. This occurred with each researcher for their own focus groups. A subsequent face-to-face meeting of the researchers discussed the individual findings and commonalities and differences were determined.

In terms of ensuring that qualitative findings 'are worth paying attention to' (Lincoln and Guba, 1985, p. 290), the trustworthiness, that is, the credibility and legitimacy of the data and the interpretation, must be ascertained. This can be done through triangulation, which in essence is the cross-checking of information and conclusions through the use of multiple processes and multiple sources of data. In this case, multiple data sources and researchers from three countries, with different youth contexts, established a framework for the corroboration of emerging themes. Instead of member-checking the researcher's interpretation of the group discussion after the conclusion of the focus groups, participants' responses were clarified during the session. Data saturation was established (when no new information was forthcoming or evident), via written responses from four classes of students from one of the schools in Adelaide, South Australia (Grades 9, 10, and 11; aged 14–16) to three open-ended questions similar to those used in the focus groups: (1) How/In what ways is technology being used by young people in positive ways in relationships with others? (2) What are the benefits for your relationships in using technology? and (3) What hinders/hampers your use of technology in relationships? These responses also served to provide triangulation with the overall focus group data. The face-to-face meeting of the researchers confirmed this process was robust and dependable.

Findings and discussion

Each country reported subtle differences and unique perspectives which emerged in terms of the positive ways that social networking and technology served their relationships within their cultural context. One overall finding across all countries is the current preference for Facebook. Each country's key themes are presented in turn, so that the breadth and depth of the views are captured.

Australia

> Facebook is positive [for our relationships] ... because it helps you interact with friends and helps you organize things in easy ways.

Young people from South Australia articulated that technology in general, and social networking sites in particular were important for their relationships as: (1) they were convenient and cheap; (2) they improved confidence; and (3) they enabled communication and social skill development; developed help-seeking behaviour and coping mechanisms; and encouraged learning and sharing of new things.

The first theme which emerged spontaneously and consistently across the three focus groups and the written responses from youth in South Australia, was that of the ease and convenience of the social networking sites, and that this positively impacted on their relationships simply because they could make contact often, for long periods of time: 'from 5–11, that's when everyone's, like ... prime time online' and with little or no cost to themselves, unlike using their mobile phones, which often they had to pay for. As will be seen below, this ease and convenience were also themes in the other two countries.

Australian participants also reported that they gained confidence as a result of using social networking: as they could 'talk to and become closer to people that you wouldn't normally talk to at school ... ' and that there was 'less chance of awkward moments occurring' and they could 'solve problems by discussing with others'. While it is generally accepted that young people use SNSs to socialize and to communicate with friends, the development of their confidence levels at an age when most feel some social insecurity and awkwardness, is of import. The Click and Connect Report (Australian Communications and Media Authority, 2009) identified that young Australians were chatting to those they already knew, but this finding, of a perceived increase in confidence, is something new and relates to the positive impact that technology is having on young people at a time when traditionally there is some social awkwardness. When taken into account with the theme which follows, it can be seen that the use of SNSs enhances not only their existing relationships, but also their own ability to manage them.

South Australian youth in this study indicated that the most important positive use that technology provided in terms of their relationships, was that it was generally enabling for them in terms of relationship maintenance: that is, it made it possible for them to do things they would not have done face-to-face due to embarrassment or fear. Young people in this pilot study articulated that using SNSs has enabled them:

- *to exercise some control:* 'You have the power to decide [what you say or do]'; 'It is easier to tell them [my friends] something [embarrassing] you wouldn't be able to [face-to-face]; it opens up communication back at school.'
- *to practise their social skills*: 'You can say to people [embarrassing] things you wouldn't be able to communicate verbally'; 'You can organize events more easily.'
- *to take time to think before sending*: 'You can actually think about what you are going to say ... not just blurt it out.'
- *to seek help*: 'a lot of us open up to our friends on the computer rather than in person'; 'Some people find it harder to express their feelings face-to-face ... they get embarrassed but they can write it online.'
- *to learn and share new things*: 'We had an assignment ... and we were all talking about it, sharing, proof reading and helping each other.'

It would seem that the added layer that social networking provides between the individuals is *enabling* their progression and skill development in relationships, helping them to grow in confidence. South Australian youth have conveyed that technology makes it possible also to easily continue or pursue relationships that might otherwise have faded. What is of import for adults to accept, then, is to recognize that technology and social networking do play an important role in young people's positive online social development and their lives generally, and then to find ways of supporting this, so as to maximize the positive impact of technology and diminish the negative.

Italy

> When you are online, you feel closer to people ... Sometimes I accept friend requests of people who do not know just to make up the numbers.

Key themes which emerged from young people's focus group discussions in Italy about the positive role of social networking and technology generally, in terms of their relationships related to: (1) power and status; (2) responsibility; (3) the perception of freedom; and (4) having access to self-support websites and greater education about social issues.

In explaining these themes, it was quite apparent that having many (500+, for example) friends, equated with popularity for these students, and was deemed positive for their relationships offline, as it demonstrated to others that one had status through this visible social presence. The notion of there being competition to have the most friends was evident, but it was also clear that they understood the difference between 'real' friends and those only on Facebook: '500–600. I know them almost all ... I can count on the fingers of one hand who I don't know. Of course I haven't a specific relationship with all of them but with almost all of them sometimes I exchanged a word.'

There was also a perception that there was a lack of parental understanding of social networking and the online environment generally. This view is not uncommon, however, unique to this cultural context, these students indicated that through social networks, *they* could adopt a more 'parental' stance to help oversee their younger siblings' interactions and behaviour in the absence of their parents, even taking some control of their internet access. This was seen as a very positive role and clearly relates to accepting responsibility and to growth in independence and identity formation as young, emerging adults. In the same way that they might watch out for their brothers and sisters in the schoolyard or playground, they were willingly taking on this role online.

Social networking was also seen positively, as a 'virtual public diary', a personal space where 'you can write and think whatever you want in freedom and without control ... but only for friends'. Clearly the issue of parental involvement on their network was not appreciated, but the recognition that the diary being public was something positive, and that it was for friends rather than self, speaks to a changing cultural context: in the past, diaries have been private, lockable and hidden, and were places where innermost thoughts could be revealed and reflected upon, without fear of rejection or ridicule. This seeming need for *public* exposure to a wide network of friends and acquaintances heralds a different youth culture from what has gone before, and may help explain why young people see technology as a medium to celebrate the public nature of their interactions and relationships, whereas adults call upon what they grew up with: a more private, considered way of relating, and hence see the public aspects of technology as threatening. This raises interesting questions about the notions of private vs public mediated spaces, the need to share with others on an exponentially increasing platform, and the controls one can choose to impose or not, to limit others viewing it.

Italian students also recognized that technology served a positive function in terms of parental monitoring and freedom: that having a cell phone, for example, enabled parents to make contact, giving the parent a sense of security and control, and giving the student a sense of freedom in being able to stay out later than if they did not have a phone:

> It is essential for emergencies, especially with what is happening today … so [if] a mother calls you ten times a day … [she] is more appreciated than one that makes five calls … they can annoy you at that moment … but then I realize that she is just worried.

Finally, there was clear recognition that social networking and technology generally have served to raise awareness of public issues: 'The network is a means of raising awareness … through the internet … [it] … can expose, discuss and build sites on anti-bullying.' Further to this, it also provides avenues for self-help, which supports the growing independence and emerging self-sufficiency of youth. If adults continue to provide offline help in conjunction with evidence-based, online websites which support young people as they learn and progress through their relationships, then we will be serving their need to engage socially online, yet also providing a safe base offline to call upon.

Denmark

> It is a kick. You can become high: What have your friends done? Has something new happened? It is like a rush through the body just before you access.

Young people in the study from Denmark articulated the following core themes relating to their relationships and SNSs: (1) reachability; (2) relationship via technology; (3) trust and safety as interrelated; and (4) a code of conduct.

This group identified that not only is Facebook their preferred SNS because of the possibility of privacy settings (vs Arto, Bebo, or MySpace) but because of its access to all their friends on one SNS. Other SNSs are viewed as either old-fashioned or childish and sharing SNS with parents and grandparents is a new experience for this group.

It is evident that reachability and relationship maintenance are core themes of central importance (Patton, 1990) as they accord with those previously reported by youth in Australia and Italy: that technology provides an easy, convenient mode of communication which enables relationships to be maintained.

One of the specific themes that the Danish youth underlined as positive and fundamental, is how Facebook is not a separate part of one's life but rather that it is totally immersed in their daily life: that the online life does not make sense *without* the offline life. Or to put it more radically: there is no distinction, because it is all part of the *same* life. As this quote highlights, Facebook-life and school life are inseparable.

> You can say that Facebook is an after-school life. We spent time together in school, but also on Facebook. And Facebook is both after and in school. And in school we talk about what happened on Facebook ... you miss out if you're not on Facebook.

However, individuals approach Facebook with more or less individual fervour, and this is evident when this group described the *emotional* state when logging on: with some describing it as a 'rush' and a getting a 'kick', while others reported logging on mostly out of boredom. What is of import, however, is that *all* reported logging on, on a daily basis and leaving Facebook open for the rest of the day. As one explains: 'I log on, leave it open and see what happens, until I go to bed.' Many of them perceive their use of Facebook to be a positive addiction. They describe it as a need to be in relation to others, to be updated and to be accessible. This is, thus, a critical element in being socially connected for today's youth.

Importantly, this group also reminded us that positive uses are simply 'the flip side of negative uses': that what is good about online access to friends is also potentially negative. You can be part of a well-established group of friends where you feel secure and belong, but the maintenance of relationships can flip into complicated dramas that are damaging to relationality.

Of particular interest is that this group highlighted that there is an unwritten code on what is acceptable behaviour online, and that they would disrespect others for breaching it: for example, the act of *tagging* was considered bad taste, and was thus perceived as not respecting an unwritten code of conduct. Also, updating content deemed 'empty' too frequently such as 'I am going to take a shower', or private affectionate exchanges between lovers or fights taking place on walls which would be responded to with comments such as 'take your fight somewhere else, you're spamming everyone's walls'. In addition, having a sensible number of friends on Facebook, those that you actually knew, gave more 'credibility', than having large numbers that you may not know. This is in direct contrast to the young people from Italy, who suggested that having more friends equates to higher status in the peer group. Finally, this group of young people pointed out that there are distinct sub-groups: some of which have to have the latest gadgets; and some of them enjoy the 'vintage' aspects of deliberately having the oldest Nokia that might even be missing four buttons. Clearly codes of acceptance and rejection operate online and these represent the two sides of the relationships coin.

Youth in this study also reported that adults' (mis)trust of them when online was an issue, but that trust and safety are tightly connected with their code of conduct. They argued that they have learned the skills to protect themselves, and that they also personally add new controls on their Facebook pages, restricting access to certain parts, for example, and suggest that adults need to give them more credit and to trust that they know how to operate safely in this environment. However, as boyd puts it: 'Teens are not necessarily well-prepared to navigate complex social worlds with invisible audiences, but neither are adults' (boyd, 2007, p. 18).

Finally, while the technology itself has agency, it is the interaction with youth who challenge the options presented by it, which generate change in the technology. Adults, then, need to acknowledge how much young people have developed as they have grown alongside technology in their lives and that 12-year-olds are different from 15-year-olds in terms of their understanding and knowledge of how to operate safely and responsibly online. Prensky's (2009) notion of digital wisdom is evident here as is boyd's caution that we need to figure out how to educate youth – and adults – in how to navigate mediated social structures and that this can best be done by learning from teens' experiences and helping them navigate via such insights (boyd, 2007 p. 22).

Conclusion

Young people have provided much needed insight into the positive ways that SNSs can influence and support their relationships and their capacity to manage and maintain them. We have seen that the risks, challenges and opportunities from being on SNSs are intertwined: that the positive uses are 'the flip side of the negative uses', that they see using SNSs as *enabling* of their communications and friendship driven practices, and that pragmatically, SNSs are cheap and convenient. Worth highlighting here is how *affectively loaded* and intertwined their online and offline lives are for them and that the lived realities of youth from this study point to the addictive mechanism of social updates.

In addition, we are reminded that being 'on' is more than merely a number of contacts. Based on this three-country pilot study, 'being on' can be specified as: accessibility, safety, trust and being part of a community. These commonalities highlight that while there are unique aspects to each country's experiences with SNSs, the youth peer context of socializing through this medium is a constant and that youth life in 2011 means more than individual subjects being online in a technologically mediated communication. It is rather a complex question of relationality, and friendship-driven practices (Ito *et al.*, 2008).

In conclusion, these insights are helpful in directing future areas of study, particularly for research carried out at the intersection of quantitative and qualitative studies. These pilot study findings thus suggest that, rather than positive uses of SNSs existing as a set of independent elements, 'intertwinement' is a key issue, which raises the following matters for future consideration.

The intertwinement of safety and trust. As adults rush to ensure that young people are safely using the online environment, what are they also doing about trust? If we think about safety as needing to take precautions and a code of conduct that can be initiated not only by governments, but by youth, how can adults and youth reconsider safety and trust as woven together: as mutually reciprocal responsibilities? *On- and offline youth life is clearly intertwined* in spectacular ways, as in cases of cyberbullying, and in the subtle ways of everyday youth life, as the youth in this study have indicated. It is evident that friendship-driven practices are the vehicle for positive and negative interactions on SNSs. Perhaps *a clear-cut distinction between cyberbullying and positive uses* of ICTs/SNSs is more blurred than previously thought, as youth in this study have clearly articulated that they are different sides of the same coin. It may be that they form the 'weft and warp' of the tapestry of online/offline relationships. Youth navigate both aspects through SNSs, and are thus also 'learning to navigate networked publics' (ibid., p. 23).

This raises the question of how we understand, and reproduce this notion in future scholarly work. Current research has shifted the focus from the offline to the online environment, viewing them as different spaces: seeking clarity about behaviour in one space versus another. Consideration needs to be given so that research does not perpetuate this separatist view of youth life. This study suggests that future research should approach investigations into youth cultures and practices in a way that respects and recognizes this online/offline 'sameness' of contemporary youth life. In closing, research questions which continue to try and separate online and offline lives may be invalid as technological developments continue to push the boundaries which draw these two elements of youth lives closer together. Youth in this study have given voice to many important, and positive aspects of the impact of technology and SNSs in particular. These insights should be celebrated and used to direct future research.

References

Australian Communications and Media Authority (2009) *Click and Connect: Young Australians' Use of Online Social Media*, available at: http://www.acma.gov.au/webwr/aba/about/recruitment/click_and_connect-02_quantitative_report.pdf (accessed May 2011).

boyd, d. (2007) 'Why youth (heart) social network sites: the role of networked publics in teenage social life', in D. Buckingham (ed.) *MacArthur Foundation Series on Digital Learning: Youth, Identity, and Digital Media*, Cambridge, MA: MIT Press.

——(2008) 'Taken out of context: American teen sociality in networked publics', PhD dissertation, University of California, Berkeley.

boyd, d. and Ellison, M.B. (2007) 'Social network sites: definition, history and scholarship', *Journal of Computer-Mediated Communication*, 13(1): 210–30.

Haavind, H. (2000) 'Staying out or getting involved: the significance of oppositional and anti-social activities among children in their coming of age', paper presented at the IIIrd Conference for Sociocultural Research, São Paulo, Brazil, July.

Harper, D. (2002) 'Talking about pictures: a case for photo elicitation', *Visual Studies*, 17(1): 13–26.

Hinduja, S. and Patchin, J.W. (2008) 'Cyberbullying: an exploratory analysis of factors related to offending and victimization', *Deviant Behaviour*, 29: 129–56.

Ito, M., Horst, H.A.S., Bittanti, M., boyd, d., Herr, B., Lange, P.G., Pascoe, C.J. *et al.* (2008) *Living and Learning with New Media: Summary of Findings from the Digital Youth Project* (the John D. and Catherine T. MacArthur Foundation Report on Digital Media and Learning), available at: http://mitpress.mit.edu/catalog/item/default.asp?ttype=2&tid=11940 (accessed July 2011).

Kofoed, J. (2004) *Elevpli. Inklusion-eksklusionsprocesser blandt børn i skolen.* [Appropriate pupilness: processes of inclusion and exclusion among children in school], PhD dissertation, The Danish University of Education, Copenhagen.

——(2009) 'Genkendelser af digital mobning, Freja vs Ronja vs Arto vs Sara vs Emma,' in J. Kofoed and D.M. Søndergaard (eds) *Bullying: Social Processes Gone Astray*, Copenhagen: Hans Reitzels Publishers.

Kofoed, J. and Ringrose, J. (2012) 'Travelling and sticky affects: exploring teens and sexualized cyberbullying through a Butlerian-Deleuzian-Guattarian lens', *Discourse: Studies in the Cultural Politics of Education*, Special Issue: Education and the Politics of Becoming.

Kofoed, J. and Søndergaard, D.M. (eds) (2009) 'Mobning. Sociale processer på afveje', [Recognition of cyberbullying.:Freja vs Ronja vs Arto vs Sara vs Emma], in J. Kofoed and D.M. Søndergaard (eds) *Bullying: Social Processes Gone Astray*, Copenhagen: Hans Reitzels Publishers.

Larsen, C.M. (2010) *Unge og online sociale netværk. En neksusanalytisk undersøgelse af medierede handlinger og offentlige diskurser*, Aalborg: Institut for Kommunikation, Aalborg Universitet.

Lenhart, A., Purcell, K., Smith, A. and Zickuhr, K. (2010) *Social Media and Young Adults*, available at: http://www.pewinternet.org/Reports/2010/Social-Media-and-Young-Adults.aspx (accessed May 2011).

Lincoln, Y.S. and Guba, E.G. (1985) *Naturalistic Inquiry*. Beverly Hills, CA: Sage.

Livingstone, S., Haddon, L., Görzig, A., and Ólafsson, K. (2011) *Risks and Safety on the Internet: The Perspective of European Children, Full Findings*. London: LSE, EU Kids Online.

Miles, M.B. and Huberman, A.M. (1994) *Qualitative Data Analysis: An Expanded Sourcebook*, 2nd edn., London: Sage.

Patton, M. (1990) *Qualitative Evaluation and Research Methods*, Beverly Hills, CA: Sage. pp. 169–86.

Prensky, M. (2001) 'Digital Natives, Digital Immigrants', On the Horizon, 9(5), available at: http://www.marcprensky.com/writing/Prensky%20-%20Digital%20Natives,%20Digital%20Immigrants%20-%20Part1.pdf (accessed January 2011).

——(2009) 'H. sapiens digital: from digital immigrants and digital natives to digital wisdom', *Innovate*, 5(3), available at: http://www.innovateonline.info/index.php?view=articleandid=705 (accessed 5 April 2010).

Rivoltella, P.C. (2006) *Digital Literacy Tools and Methodologies for the Information Society*, Hershey, PA: IGI Publishing Group.

Save the Children (2008) *Ragazzi connessi. I pre-adolescenti italiani e i nuovi media*, December, available at: http://www.savethechildren.it/IT/Tool/Pubblicazioni/All?year=2008 (accessed May 2011).

Smith, J.A. and Eatough, V. (2007) 'Interpretative phenomenological analysis', in E. Lyons and A. Coyle (eds) *Analysing Qualitative Data in Psychology*, London: Sage.

Spears, B.A., Slee, P.T., Owens, L. and Johnson, B. (2008) *Behind the Scenes: Insights into the Human Dimension of Covert Bullying*, report prepared for DEEWR, Canberra.

——(2009) 'Behind the scenes and screens: insights into the human dimension of covert and cyberbullying', *Zeitschrift für Psychologie* [Journal of Psychology], 217: 189–96.

Stauæs, D. (2004) *Etnicitet, køn og skoleliv* [Ethnicity, gender and school life], Copenhagen: Samfundslitteratur.

Storey, L. (2007) 'Doing interpretative phenomenological analysis', in E. Lyons and A. Coyle (eds) *Analysing Qualitative Data in Psychology*, London: Sage.

West, L. (2004) The learner's voice: making space? Challenging space? Keynote address, 2004, Canterbury Action Research Network (CANTARNET) conference, reported in *The Enquirer*, Spring 2005. Available at: http://www.nya.org.uk/hearbyright (accessed May 2011).

Williams, K.R. and Guerra, N.G. (2007) 'Prevalence and predictors of internet bullying', *Journal of Adolescents' Health*, 41: 814–21.

Chapter 2

Secure net addresses

Secure internet and responsibility

Haukur Arnthórsson

National net addresses comprise a system which formalizes people's identity on the net through people using their own names. Certain bodies, such as the government and political institutions and even businesses, need to be able to ensure someone's identity is valid. This solution does not affect the anonymity of the net, but can work alongside it.

National net addresses are able to considerably increase the security of the net and this fact therefore makes them very suitable for children, youths or the elderly and it is natural for schools also to make full use of it. National net addresses combine American identity solutions, which are built on net addresses, and European solutions built on certifications. Together they make a system compatible in both markets.

The information revolution is subject to complex laws, including marketing laws, but as people have created it, it can be changed. To the benefit of the broader public and the business community, many Western politicians have begun to be interested in information technology issues. In addition, the status of smaller states and the Third World has become part of the discussion in relation to large global companies. Most of these global companies have their headquarters in the United States, and lead the development of the internet for the global community. In contrast, China, however, is leading development aimed at the residents of each state, ensuring services on the internet are provided by domestic parties.

At present it is hoped that information technology will give rise to a new, improved society. The internet is an important part of modern life, especially for young people, children and teenagers. However, anonymity for the public on the internet creates considerable danger for some, and there is a real need to give greater priority to social and political goals in this discussion.

Currently, there is a great deal of discussion concerning security issues and access to the internet and this history stretches back more than 30 years. Concepts such as *net identity* and *digital certification of users* are often referred to. The focus of the discussion over time has gradually shifted away from technical solutions about security, and has moved steadily closer towards the user side. It can be asked whether, in this respect, this development is complete. Early discussions emphasized the certification system, with the Public Key Infrastructure (PKI) system emerging in 1976. About 10 years ago the emphasis shifted to *smart cards*. Then the Finnish government distributed plastic cards with microprocessors to the entire Finnish nation. This measure was not a success for many reasons, for example, the lack of the necessary infrastructure, as it emerged that there was another and greater need for digital certification for internet communications to be adopted throughout the entire society. Recently, in some quarters, especially in discussions in North America, emphasis has shifted to *net identity*.

Presented here for discussion is the notion of net identity, which is connected to a personal identification number (SSN) in a national registry. Building upon all previous technological innovations, this idea further develops the concept as the user's correct name is now required to be part of the information provided when connecting to the internet. Such a net identity is referred to here as a secure net address. Such net addresses could greatly increase security on the internet and are likely to have a significant social impact if successful.

Several concepts are relevant to this discussion, and their contexts vary in complexity, for example, *net identity* is very nearly the same as net address (everyone knows about e-mail addresses); *certification*, which is confirmation of who is who and is usually based on a certification system. For entire nations that system can be the open Public Key Infrastructure (PKI), but otherwise systems of private parties, such as VeriSign, exist. For certification to be implemented, an internet user has to supply confirmation of who he is in respect to the computer on which he is working, and this is also called *identity*; fingerprint and voice identification are both forms of *bio-identity*; however, some trust providers also use smart cards, which are called *digital certificates*. All of these units and concepts work together with authentication in a coordinated access framework.

The current situation

Anonymity is the main rule of the internet. Until recently identity management was completely distributed on the internet, which means that each and every user has a unique username or net identity on each and every Internet Service Provider (ISP). Thus, users can have many net identities and provide different information about themselves with different service providers and even with the same one. Likewise, people can present themselves under these usernames on different occasions.

Centralized net identity

Several years ago, a new identity service emerged that was centralized (see www.OpenId.com). The public can register at the website, provide information about themselves and, after registering, people do not have to repeat this process and enter their identity elsewhere on the internet. OpenId is an identity server, that is, other websites that a user visits are supposed to retrieve the user's identity and information from OpenId. This system has now become widespread, and people all over the world use it.

Nevertheless, the security level of this service is deemed insufficiently secure for administration in the United States. The Obama administration has subsequently entered into an agreement with OpenId and has formed a new basis for the public's official communications. The change entails that an identity from OpenId and other parties is certified by a trust provider.

The American solution will provide a centralized, standardized alternative, although implementing it will involve many parties. The websites visited by an internet user must retrieve an identity for and information about the user from a centralized location on the internet: an identity server. The server will guarantee the public's net identities in respect of other parties on the internet for the individuals involved (OIX, 2010). The government and marketing groups in the United States recently founded OIX (www.OpenIdentityexchange.org) (March, 2010), which should increase trust in communications by using net identities for government and other parties for shopping and business transactions. The US Government has agreed that the agency will certify the security of identity servers operating in accordance with government standards. Google, PayPal and Equifax are the first identity servers to receive certification from OIX. They certify the net identities of individuals logging onto (government) websites (ibid.).

Proposal: Iceland

Iceland enjoys a unique status as it is situated halfway between two continents and engages in business with both of them. Let's imagine that we want to use the same system for American and European parties visiting websites as for a website of the Icelandic administrative system. We then have to support the main characteristics of the American and European approaches as the system's basis. An identity service provides interoperability with the American solution, while a trust provider provides this for the Nordic/European system.

Digital certificate is being standardized in Europe. This means that those people trading with banks have a digital certificate from the bank and have a card reader on their computer that can identify them with extremely high security. However, others with no certificate, such as children, youths or elderly people, foreign citizens with a residency permit and all others not trading with a bank, will connect to the internet via a username and password. Their usernames are nevertheless certified, just as in the American system. This is an important difference. Trust providers will be able to certify a national net address with a high level of security, for example, for voting, if the public has certificates for the purpose, but at a low level of security if a username and password are used.

The proposal is: *The entire Icelandic nation and others in the National Registry are allocated or choose a national net address, which is unique to them.*

National ID and national net address

National net addresses would become stand-ins for personal identification numbers for access management and the dialogue appearing when connecting to a website. A personal identification number will not be a net address.

Note that a government website visited by the general public can access all information about the user from the identity server, not only a net address but also more detailed information, if requested. For example, a government website can request the name, personal identification number, address and date of birth of the user. The internet user would have to approve the provision of this information and it is very important that the user determines what information is provided. This would therefore improve the present protection of privacy if this approach were adopted in Iceland.

The personal identification number would otherwise continue as an identity for the public, but the national net address might change, and individuals could choose them at their discretion. This pin number will not be official and will not be recorded in a national registry.

Centralized management of net identities

Now let us turn to Iceland and the Nordic countries. In Iceland, in cooperation with banks and the government, preparation of a certification system, based on digital certificates in credit cards, using microprocessors, is underway at present. This system is well suited to banks and provides an extremely high level of security. This system is comparable to one that has been tried in Finland and Estonia.

The system involves communication solely with a certification server. It can be connected to e-mail addresses, including any existing e-mail address, and can, for example, certify e-mail as if a business transaction were involved. The system assumes distributed management of net identities but is not centralized. This system is not connected with the American approach.

The status of centralized net identities as a solution to access to internet services was strengthened with the agreement of the US Government. Both perspectives of distributed or centralized solutions for net identities were competing for the right to operate when net identities achieved this new facility.

Discussion

National net addresses would become net identities that the public could use for official communications on the internet instead of using net identities from their workplace or from single service providers. They can be used for diverse purposes, including connecting to official websites, government social media, responsible mass media and commercial websites.

The innovation involves the merging of ideas developed in North America about 'trusted federated identity management', the PKI (Public Key Infrastructure) system and the internet ideas on identity transfer between parties, such as those presented by the OpenId collaboration, together with ideas from Nordic ministries of justice about national net addresses with a private e-mail library. Ideas of this kind have not actually been implemented anywhere yet. They entail ethical, cultural and social changes that will have a great impact on the internet. They can cover an entire nation or a particular roster of names, whether of associations or companies.

With their advent, government parties can standardize and control access to their systems without entering into agreements with Icelandic workplaces or international e-mail service providers. Digital certification on the internet affords an opportunity to address certain issues that otherwise would not be attempted. However, this is within the framework set for trust providers by technology and the large marketing companies. The system's main goals are to increase the efficiency and competitiveness of Icelandic society, promote environmental protection and sustainability in digital communications, and enhance security on the internet.

It is assumed that national net addresses can fulfil these main goals:

- social goals
- political goals
- commercial goals.

Social goals of national net addresses include helping children, youth and minority groups to have secure communication on the internet. It is the internet's anonymity that is very difficult for these groups, and this creates unsuitable conditions for bringing up children and youth who are in the process of learning how human relations should be performed. Bullying on the internet is the main manifestation of this issue. Anonymity also creates an inappropriate environment for elderly people, and the internet can also be used for attacks on minority groups. Anonymity can also facilitate personal attacks on the internet generally and can ruin public discussion, for example, by publishing inappropriate opinions and inciting vigilante acts. The proposal is supposed to be used by schools for communications. This could become the general means of communication for children and youths in Iceland. Parents can then permit or teach their children to use anonymous e-mail addresses, but they would do so at their own risk and without government support.

Specific *political goals* are making politics and government administration more secure and substantive. Again, anonymity is an enemy that cannot be tolerated. Generally, politics and government administration, in particular, cannot function without knowing who they are dealing with. This applies equally to government services as simple as dispersing child benefits and as complicated as participation in electronic voting. Following discussion of the need for bloggers and mass media to become more responsible than they have been, and in light of the increased role of direct democracy on the internet in our societies, stronger internet constraints are required. Appearing under an identifiable name could be a step in this direction.

The proposal achieves its *commercial goals* as the entire Icelandic nation enters into internet transactions with independent market parties operating internet services. Their basic service is free, but the public will be able to purchase additional services. The goal of the system is to address the market deficiency that the internet services in Iceland are currently tied to domestic workplaces and international e-mail giants like gmail and hotmail and not to independent service providers in the Icelandic market. This can be done by establishing an Icelandic centre with national net addresses. It can also be called a commercial goal if the Icelandic certification system conforms to both the American and European certification standards.

Iceland has a special status since a general market for internet services has not yet developed. The majority of the Icelandic public therefore uses their workplace e-mail addresses. Government parties and banks, as well as others working to support a paperless society, cannot send letters to people at their public e-mail addresses at workplaces; people need to have e-mail addresses that are not read by other employees. Icelanders not using a workplace e-mail address most often use e-mail addresses at gmail or hotmail. This can be viewed as a market deficiency in Iceland as the public generally does not have e-mail addresses supplied by an independent Icelandic party. The state can fix this by allowing the public to choose national net addresses.

New markets on the internet

A national net address offers the new possibilities for services from domestic bodies provided to individuals on the internet, such as data storage. Additional use is inconceivable if the personal identification number is used as a net identity because the personal identification number is such a sensitive identifier. With national net addresses a new option can emerge in an open market on the internet for Icelandic service providers.

National net addresses would be good for schools and other government bodies, where it is normal to use a name. Also, a national net address could open new avenues for services for the self-employed and for small companies; most companies in Iceland come into this category.

By implementing national net identity on the internet in a centralized system, agencies and companies can adopt certified communications without adapting their own equipment because the same system will receive and deliver certified mail. Thus, government bodies can demand that mail to them be certified and can make other databases completely digital without incurring separate costs. This is important in times of economic constraint. The implementation of digital certification, on the other hand, calls for changes to the internal systems of agencies and companies.

Social media

With the advent of social media, demands to appear under a name have increased. Government bodies as well as municipalities or schools can set up their own websites for communication and consultation with the people. It is also important for these websites to be open to only the right parties, for example, so that the equipment is not used for dubious purposes. The personal identification number is the key to providing and limiting access to social media, and it may not be used directly, that is the function of the national net address.

Can a national market be created?

Increasingly China has gone its own way in providing competition to the international and American systems and internet solutions, and is leading small nations and the Third World in this regard. Iceland can also do this. With national net addresses a certain market will doubtless be transferred to the public domain.

Bio-identity of the future

In the future, users will use their telephones, and after establishing who is speaking, the party will be certified – and then connection with the requested service will be obtained – on the same telephone. A bio-identity such as voice recognition, which is still at the developmental stage, fits in very well with the development of the public using their telephones rather than their computers as the main computer medium in the future, with everyone appearing under his or her own name.

Conclusion

The justification for launching the project discussed here comes from its social impact. It primarily provides everyone, including children, youths and elderly people, with a net address from an independent party, not subject to commercial ties. It also opens new social possibilities and new markets and strengthens responsible use of the internet. National net addresses could be a normal government intervention in information technology.

Acknowledgements

The author has received a grant from the Icelandic Centre for Research and is working on the concept of national net addresses.

Reference

OIX (2010) *An Open Market Solution for Online Identity Assurance*, press release, available at: http://OpenIdentityexchange.org/press-releases/oix-launch-2010-03-03 (accessed 3 March 2010).

Chapter 3

Media education

A new academic curriculum

Enrico de Santo and Angela Costabile

This last century of achievement can be considered the 'mass media century': Lumière's first cinematographic projections, the birth and the development of radio and television followed by the emergence of the internet can be considered the most notable events. Mass media seem to have underlined a turning point in communicative and technological developments. The significant change in communication dynamics and, moreover, in cultural sets caused by the huge advent of mass media, has greatly contributed to the shift of interests in social studies of communication. In the past 30 years, the interests of social scientists have focused on the consequences of the cognitive, perceptive and representational activity of individuals caused by the extended exposure to mass media, in other words, the social construction processes of life focusing on the communication–education nexus. It is emphasized that in technologically developed societies, young people grow up in a completely different social environment, where multimedia experiences can be considered normal everyday life.

Apart from being a typical entertainment experience, mass media are an extraordinary source of knowledge for young people, even if sometimes this aspect can be seen to be negative, as it develops excessive self-learning dynamics, e.g. learning by doing, causing the loss of the typical guide-function-to-knowledge of mass media (Greco, 2004).

Moreover, mass media act as socialization carriers for their cognitive-emotional contents whose modalities can be considered deeply influential, more so than family and school, thus evidencing a deep crisis of moral values and communicative incapability.

We find ourselves, unlike in the past, in a completely different reality of rapid acceleration and unceasing checks of the power of socialization, in which friendship groups and mass media act as universally open elements (Morcellini, 1997).

The actual environment appears open to new perspectives (new democratic and creative opportunities of knowledge and socialization) but also undeniably open to risks (anti-socialization processes and new inequalities). All these points create an educational revival owing to incessant shifts caused by the transformation processes of the mass media (Greco, 2004).

Although studies of the relationship between mass media and young peple show different tendencies, aside from the pessimistic (apocalyptic) views that blame mass media for educational failures, we can list some issues regarding young people and their ability to be included in social contexts through the mass media as follows:

- We can claim that mass media are 'poor teachers'.
- We can claim that social values, fragmented and enacted by mass media create a stimulus overflow, which in some cases could be contradictory, generating confusion and upset.
- We can emphasize that the use of mass media is surely more entertaining than reading a book or something else.

From this point of view, emphasizing those issues, building a new kind of education is indispensable, particularly in the context of the decline of all forms of social mediation, and with the introduction of changes made by media-morphosis processes, which act as new forms of socialization, knowledge and communication (Greco, 2000).

As social systems develop from a media point of view, the educational system must supply media skills, i.e. skills in understanding mass media, or media competencies, i.e. the ability to create new forms of expression and communication. This assumes mass media are not only neutral carriers of knowledge but are a new 'way of life', building individualistic socio-relational and cognitive-emotive experiences. In this sense, a specific 'making of knowledge' for those who act in educative contexts is necessary, suggesting the development of specific skills for the future direction of what a society requires.

Moreover, the classic pedagogical point of view of 'codified knowledge', intended as a knowledge flow from master to pupil, seems inadequate now. It is necessary to verify if a new system based on mass media and the traditional way of teaching could promote a more efficient system of knowledge and a change in the relationship between master and pupil (Cesareo and Rodi, 1996).

We should say that the skill of being a part classic 'master', and part 'street guy' is necessary for a teacher, which is to say that an expert teacher is needed whose interventions are directed to the selection of modalities and stages of self-learning; an interlocutor whose target is 'learning by doing', and who is accepted by the pupil as someone who builds something together with him and is not only a mere 'knowledge transmitter' (Morcellini, 2000). According to Piromallo Gambardella (1997), a new form of education is needed, one that is new and personalized, requiring a specialized educator and a different setting based on 'community'.

According to ibid., the creation of a media education-based educative system could be suitable. That is, the creation of a critical sense throughout the mass media. In this sense, a coherent answer could be found in the media education project (ME). The term 'media education' was first used by Len Masterman to point out several experiences worldwide that compare the 'new' education with the 'old' education (Giannatelli, 1996).

Media education aims to direct government attention to the importance of educating individuals from school age through to adulthood, to encourage a positive relationship with the mass media (Rivoltella and Marazzi, 2001). The Anglo-Saxon term 'media education' expresses, better than other neo-Latin terms, the connection between mass media and the education world. This term emphasizes three main aims:

1 An education *in* mass media, that is, to a critical point of view as regards mass media messages system.
2 An education *with* mass media, that is, in the use of mass media as important means for education.
3 An education *for* mass media, that is the creation of a new professional role (media educator) acting for new educational targets.

Media education's target is not directed at the 'defence' of mass media, that is, to protect individuals from mass media effects, but surely is an 'attack strategy' in order to enable new skills in understanding mass media dynamics.

History of media education

The media education evolution began in the 1970s, thanks to UNESCO's patronage, spreading it worldwide. Early signs can be found, however, in the 1940s. In Italy, the Reverend Felix Morlion introduced the 'cineforum' (the debate after a film) strategy, defining it as 'one of the means of attracting and of uniting men and women from all social classes'. Following this experience, in the 1950s, the Centro Studi Cinematografici (CSC) launched a plan to promulgate critical interpretation of cinematographic image through a specialized 'cineforum'. This experimentation developed, during the 1960s, into the experimentation of cinema teaching in schools. In Great Britain at the end of the 1950s, the Newson Report underlined the importance of 'studying cinema and television and the need to train pupils to understand critically what they see' (Rivoltella, 2001).

In 1967, in Lausanne, a politician suggested integrating teaching of *Cinéma et Language* in primary and secondary schools. This enterprise was the foundation for the Centre d'Initiation et aux Moyens de Communication de Masse (CIC), a governing body with the aim of supporting mass media experiences in schools. Four years later, the pastoral institution *Communio et Progressio* underlined the importance of media education in the church's thought:

> The Church thinks that a Christian education for receptors is strictly necessary ... the organization of Catholic schools must face more

> seriously this duty in this field; in every school it has to enact a teaching that forms pupils not only as readers, listeners or viewers but also giving the opportunity to use actively all the possibilities of expression that all mass media can offer.
>
> (Giannatelli, 1996)

In 1973, we find the first official definition of the term 'media education'. It is included in a document of the Conseil International du Cinéma et de la Télévision (CICT), a UNESCO organization. In this document, media education is defined as:

> The study, teaching and learning of modern mass media as means of expression, considered as specific and autonomous subjects in the context of an educational practice, in opposition to the use of mass media as teaching aids for knowledge, such as math, science and geography.
>
> (Pavlic, 1987)

The 1970s started with a new evolution in media education: Australia was the first nation to adopt media education in its scholastic curriculum. In 1978, the *Association for Media Literacy* (AML) was established in Canada, an association consisting of parents, teachers, and media professionals. In the 1980s, John Pungente, a Jesuit Father, undertook a survey on teaching media education in 400 Jesuit secondary schools all over the world. This survey aimed to explore three main questions: (1) What is the state of art of Media Education in Jesuit schools? (2) What is the state of art of Media Education in the schools all over the world? and (3) What can be done to encourage Media Education in Jesuit schools?

After visiting Jesuit schools across 29 countries and 62 cities, Pungente compiled a report dealing with the state of the art of Media Education, outlining five key areas: (1) What is Media Education nowadays? (2) Why is Media Education important? (3) How can we answer objections to Media Education? (4) What are the patterns for teaching Media Education? and (5) Providing an essential bibliography for media educators.

In the same year, Len Masterman published his book, *Teaching the Media* (1985), in which a deeper response to John Pungente's questions can be found: (1) What is Media Education? (2) Why it is so important? and (3) What can be done for state teachers performing Media Education?

In 1983, in France, CLEMI (Centre de Liaison de l'Enseignement et des Moyens d'Information) was born, a special French governing agency whose aim is to encourage, through teaching actions, multiple use of mass media in teaching, in order to promote in pupils the best comprehension of their world, permitting them to develop a critical perspective for mass media. In Italy, Media Education follows the Anglo-Saxon tradition, particularly aligned with Len Masterman and John Pungente. The first approaches to Media

Education can be traced to the early 1990s when Roberto Giannatelli, one of the Italian Media Education founders, began his first Media Education workshops in 1991 at Istituto di Scienze della Comunicazione Sociale (ISCOS), which was founded in 1988, in the Università Pontificia Salesiana in Rome. In February 1996, the Italian Association for Media Education and Communication (MED) was established and, recently, the Corvara Summer School.

Media Education approaches

To understand what Media Education can represent today and what it could become in an educative context, it is important to outline the various approaches that this branch of knowledge has taken during its history. Let's start with the 'inoculative' approach, also called the moral approach. From the 1930s to the 1950s mass media were considered to be corrupters, a sort of illness menacing the 'cultural and moral health of man' due to their influence in inciting consumerism (Buckingham, 2003). In other words, mass media were considered a 'virus' and young people must be protected from it: this explains the term 'inoculative'. In this sense, young people must be protected from 'forbidden' contents that could damage their physical and mental health. In this sense, Merton's approach is very important. According to Merton (1957), mass media can be 'dysfunctional', that is to say, can create disharmony, stimulating deviant behaviour in young people. According to this point of view, the choice was between an education where the difference lay between 'high' culture, and 'low' culture, that is to say, moral values vs popular ones. Those interventions were considered an 'antivirus', a vaccine protecting pupils from the bad influences of mass media. The inoculative approach can be traced to some theoretical communication models, one of which is the Magic Bullet Theory. According to this theory, mass media stimulate media reception in a passive listener. In the 1950s and 1960s, the Popular Arts or Critical Reading approach was considered. This approach emerged from Stuart Hall and Paddy Whannel's (1967) book, *The Popular Arts*, containing some suggestions for teaching about media with particular reference to cinema.

In the Popular Arts approach, mass media cultural contexts are revalued, particularly those of cinema: mass media becomes the people's expression in spite of being an ideological apparatus. Culture became the main subject of scientific research, especially in Great Britain when the Centre for Contemporary Cultural Studies (CCCS) was established in Birmingham. Those studies then demonstrated how all expressions of popular culture are worthy of consideration. However, this approach failed in some respects, as outlined by Len Masterman. According to Masterman, this method was wrong because:

1 It was paternalistic and protectionistic as media educators intended to impose their preferences.

2 It was discriminating as it privileged upper-class tastes and it selected and exalted some mass media to the detriment of others.
3 It was strictly evaluative, with uncertain criteria, because media educators were not able to demonstrate why a certain film could be considered 'superior' to another.

In other words, communication and mass media began to be considered carriers of culture and values of everyday life. According to this point of view, education was intended as a construction of a different way of interpreting reality of life. Media Education then became an ethical commitment assuring minimal strategies to propagate mass media critical knowledge in a media world.

Another important approach worthy of esteem is the Symbolic one, also called the Representation approach, born in the 1970s and 1980s. This approach is based on Roland Barthes' book, *Mythologies* (1957), in which the difference between 'high' and 'low' culture is described. Barthes outlined the difference between 'high culture' and 'low culture' (even a comic is 'culture') and introduced the concept of mass media 'non-transparency': where mass media pretended to represent reality but 'they are not reality'.

In fact, they are a representation of real, a 'construction' obeying particular linguistic rules and ideological interests. The representation paradigm mirrors the outline of a reticular society that can be explored according various points of view. Mass media are considered symbolic systems of reality representation, as carriers of meanings and cultural values able to impress individual knowledge processes.

However, the vision of a critical and self-governing audience, able to discern mass media's artificial making of news should be considered: mass media are supporters of the ideologies of the ruling classes. In these years, education seems to be directed through knowledge processes rather than content. Mass media were outlined as 'representations' of ideological and economic interests and this shows Media Education's trends in the 1970s and 1980s: an investigation of mass media to create critical and self-governing researchers as free citizens.

The concept of 'representation' becomes central in Media Education: it gives a theoretical foundation to what schools will do. The media educator's main task consists of giving pupils the right tools to 'deconstruct' mass media's text. Encoding–decoding activities are fundamental in Media Education. Semiotic and cultural studies reveal the methods and tools for text analysis in schools (Cappello, 1995). This last paradigm can be considered the most important if compared with others and it is nowadays accepted in Media Education.

Media Education and educational training

One of the most privileged fields of action of Media Education is the school, even if is not the only one. The school is a fundamental institution for education: and for this reason it is possible to undertake Media Education in real terms, using pedagogies that stimulate critical enquiry.

Andrew Hart, director of the Media Education Centre at Southampton University states that Media Education can be fully implemented in schools on condition that it can answer these three questions (Hart, 1998):

1 Why should we teach mass media?
2 What can be taught in mass media?
3 How can we teach mass media?

Media educators in the Anglo-Saxon tradition have answered Hart's questions as follows:

Why should we teach mass media?

Len Masterman had shown in his book *Teaching the Media* (1985) seven reasons for the introduction of Media Education in schools.

The first reason is that of mass media's saturation or ubiquity. According to the ALM (Association for Media Literacy), mass media takes up a huge part of young Canadians' lives: 11,000 hours per year are spent in classrooms in comparison to 15,000 hours in watching television.

A second reason can be found in the fact that mass media are authentic '*awareness industries*'. They are not neutral in the fact that they manipulate ways of being. A great part of young people's way of thinking and behaving is due to the influence of mass media culture. Moreover, mass media are '*news factories*' according to *agenda setting* rules where mass media decide what is important to be known by an audience or as *the gate keeper* where mass media are the 'guardians', filtering news for the audience.

There is a close relationship between Media Education and democracy. At the UNESCO Convention, in Toulouse, in 1990, the former Minister of Education Jospin closed the meeting with the following statement:

> There is not democracy without participation, there is not active citizenship without education, there is not education without information, culture, and critical awareness. If we want mass media as acting for democracy, we must start from a democratic and educative approach to mass media. School is necessary.
>
> (Bazalgette *et al.*, 1992)

We have also to consider the importance of *audio-visual media* in everyday life. Our society can be defined as an 'image society' where we live in an unbroken flux of sounds and images. In particular, young people feel the charm of audio-visual communications. This is a language that must be decoded and used even at school.

Another reason for teaching mass media in schools can be traced to the proliferation of mass media privatization: self-media, new media, and the internet. Young people's bedrooms are like a small communication centre beaming out to the world. And, finally, there is the need to educate young people for the future: in this case, the future belongs to the world of communication and to media communication (Mattelart, 1998; Thompson, 1998).

What can be taught in mass media?

Mass media teaching had been undertaken in the 1980s in advanced educational systems, even if in a non-systematic way. Image education, cinema education, press education were taught, but every medium was considered 'as is', standalone.

Since 1983, in France, CLEMI has organized '*la semaine de la presse*', when schools are involved, together with reporters, in critical reading of newspapers (Gonnet, 1995). Even Masterman suggested his book '*Teaching about television*' in a lecture about Media Education.

Masterman, in his book, *Teaching the Media* (1985), intended to create a kind of common language, a sort of *koinè*, that must be learned in order to understand mass media language: in other words, a kind of 'common culture' of Media Education. First of all, it is necessary to understand and analyze the meaning of 'representation': mass media are not reality, but they represent it; they are not 'windows on the world' but 'faked constructions' made by ideological, political and economic interests according to a complex process of news selection; they are not transparent but opaque.

Len Masterman suggests four areas of exploration to help pupils to understand mass media processes: (1) *determinants* or mass media institutions (who communicates what and for which aim); (2) *rhetorics* (their language, and techniques); (3) *ideology* (ideas, values, interests); and (4) *audience* (who gets the message, how they understand it and how this implicates another one).

How can we teach mass media?

After having characterized a subject of study and contents, it seems necessary to understand how to 'teach with media'. According to Roberto Giannatelli, after ten years of Media Education experimentation in Italy, some 'constants' can be traced which can be defined as 'method's principles', at least in regards to Italian Media Education:

- Media Education must be entrusted to educators acting in school. Only exceptionally will an external expert be preferred. It is not necessary, at present, to create Media Education as a new scholastic matter. It is more useful to create interdisciplinary teaching whose coordinator will be the teacher of Italian together with other teachers. Media Education must be, as Masterman suggests, a *'curriculum across the curriculum'*.

- Media Education will be projected as a progressive and spiral-shaped teaching, acting from a single medium through increasing progression and difficulty: from photography to television, newspaper, cinema for example.
- Production and analysis activities will be on the same level. Practical work with mass media will only be a didactic exercise. However, a teacher as a 'facilitator of knowledge' together with classroom and colleagues' manager is strictly necessary.
- Media Education must have an educative style, which is democratic and liberal, based on participation and 'generational exchange' (Scurati and Caimi, 1991). Concerning the mass media, teachers (and adults in general) can learn a lot from their pupils; but they can also teach a lot regarding culture, values, and life experience.
- Media Education will be a 'lifelong process' of education, even with adult education. If Media Education's main aim is to achieve 'critical independence' (Masterman, 1985) and 'user's competence' (Baacke, 1997), we can understand that those aims can be reached only over time and through different experiences and situations. The school can become a suitable location for Media Education.

Other fields of intervention in Media Education

Apart from the scholastic field, Media Education can be applied to profit and non-profit domains, intended for profit and social enterprise. Globalization and new media are now strategic assets of corporations. Virtual reality is now the main product that an international market requests. According to this perspective, enterprises have acquired new organizational levels. In fact, in order to be able to offer virtual products and services, it is strictly necessary for 'a widespread net of information about markets, customers, competitors be supported by new technologies, allowing enterprises an immediate adaptability of its mission to new market requests' (Rivoltella and Marazzi, 2001).

As for the non-profit field, this context aims to provide cooperation and support, as the creation of basis for the development of collaborative context. Social enterprise's main aims are the integration of those who are socially disadvantaged, the production of services connected to social needs and work perspectives for those who are socially disadvantaged.

With regard to social disadvantage, Media Education can offer tools for social diagnosis and therapeutic intervention, self-improvement, support for social relationships and communication through theatrical and musical activities. In this way, organized co-operation must be intended as a social enterprise and supported as a business one because it has both a specific target and a social significance in the modern social context (ibid.).

Media Education: a new professional perspective

In our modern society, where education and training take place in a media-enriched society, the new teacher and educator will have to be a Media Educator. But there is an important question about this figure regarding training, as it can be undertaken in three different sites: school, profit and not-for-profit arenas.

Who is a media educator? Is there a definite professional identity? While we will try to give an answer, we can begin by revealing who has been a media educator in the past, that is, a teacher or a skilled media person who worked in the school. In the former case, this identity is confused with the professionalism of a teacher: a teacher who has media skills and wants to teach students Media Education may not feel like a media educator, but is considered as a teacher who is skilled in modern societal communications and its culture and change. In the latter case, an external skilled person who makes an offer to train others could be considered to be a media educator. So what are the prerequisites of a media educator?

Rivoltella says that whatever the scope in which a media educator works, he or she must respect two logics: a systemic logic, where the organization in which the media educator works acknowledges Media Education; and an ecologic logic that integrates all media in relation to needs and contexts.

So a media educator can be different figures at the same time and have two different prospects. Let's take an example of the school. The first consideration is the teacher who has to give lessons in Media Education. The main task is training: teaching teachers how to adapt Media Education in their classes. This prospect involves the principle of the search: that is, a search is made every day and directly in the class to find the right strategy to teach. A second prospect recognizes the competence of an external skilled person who is not on the school staff. It is not possible to train teachers in the methodologies of mass media because they have a very different cultural education, so, the media educator becomes a 'fellow traveller' of the teacher. But there is still one important point to define: the training of the professional figure of a media educator.

Although the training courses for a media educator involve a degree in Media Education or a Master's in Media Education, it is necessary to consider three important points:

1 A media educator is the teacher/trainer of new generation. In a society governed by information and technologies of communication, it is not possible to choose to be a media educator, but each person that works in any instructive context has to possess media skills. This is a development in the field of the training professions: the media educator is the trainer of a new model, a teacher of a new model, a professional trainer of a new model.
2 There are three fields which can professionalize this figure: school, the for-profit and the not-for-profit sectors. So the media educator of a

company will have technology and organization skills, the social media educator will have psychological and pedagogic skills, and the teacher media educator will have cultural skills.

3 If we think of Media Education as a whole range of skills, we should not think that it will be impossible to find a media educator with a strong identity, on the contrary, it is possible that a professional figure with a strong communicative-educational expertise already exists.

A new Media Education

The new media scene has occurred through the fast evolution of digital technology and has brought about a change both in communicative processes and in educational processes. What had characterized the construction of the cultural and meaning industry now must change and search for a new definition of Media Education. Portability is the first aspect of new media. A developing trend of technology concerns the release of the user from the necessity of having a fixed position related to information: consider some years ago when the cellular phone was only a phone, and the palm-sized computer was only a personal computer. The first palmtop allowed access to the internet, even though rather slowly. Today with the fusion of these two devices, we have the smart phone. Thanks to this new system we can surf the internet, we can take pictures, film videos and send them, we can watch television on our mobiles.

This feature challenges us with new ways to use traditional media, on one hand, but, on the other, nullifies the educational processes offered by the use and culture of books. So teenagers can surf the internet without adult control and be the only one responsible for what they do.

A second aspect concerns the personalization of the new media. Digital media are personal because they belong to the private life of the subject: technology is an essential part of our life and our memories. Digital media are 'personal': their aim is interaction but it is used by a single person.

Some scholars have referred to the bedroom culture, where teenagers 'live' the personalization of technology to escape from adult control while the adult thinks that it is right not to violate the private world of their children. So technology might push the generations away from each other, but it might also reach them as a cultural negation between one generation and the other.

A third aspect concerns the multitasking configuration of new generations: let's think about what teenagers do while they are doing their homework: on their desk we can find a book, their computer, their cellular phone, their iPod. They are studying but at the same time they are using the computer to organize a text starting with what they are reading, they are chatting with a schoolfellow, they are sending sms, they are listening to the music. So thanks to new digital technology, the new generation can do a lot of activities at the same time, but this involves a development of their skills. Is that really necessary?

But there is one other aspect to consider: teenagers produce media. YouTube and Facebook are the proof of that. They show themselves to the whole world through personal videos and photos. We hear from news services, that videos of bullying are being filmed with their mobile phone, and they are posting them on the internet as if it were something to be proud of. At this point, the bedroom culture becomes the pocket culture as they take their digital world out with them.

So, now we have to think about a new Media Education that considers two aspects: an organizational and a cultural aspect. On the cultural plane the scene has to have the following features: horizontal socialization, cancellation of past time, loss of future, overemphasis of present time to give importance to emotional and tactile dimensions.

On the organizational plane, Media Education not only has to support teaching methods, but also has to migrate into the didactic, into educational intervention and become, in this way, Media Education for all.

In this way, as stated above, Media Education can be considered a concrete opportunity to develop good practices in educational contexts. Media Education can also improve teaching strategies for young people through the development of critical awareness about mass media.

References

Baacke, D. (1997) *Medienpädagogik*, Tübingen: Niemeyer.

Barthes, R. (1957) *Mythologies*, Paris: Editions du Seuil.

Bazalgette, C., Bevort, E. and Savino, J. (1992) *New Directions: Media Education Worldwide*, London: British Film Institute-Clemi-UNESCO.

Buckingham, D. (2003) *Media Education*, Trento: Erickson.

——(2007) *Media Education. Alfabetizzazione, apprendimento e cultura contemporanea*, Trento: Erickson.

Cappello, G. (1995) 'Mass media, pedagogia, curriculum. Tra riflessione teoricae applicazione pratica', in R. Giannatelli and P.C. Rivoltella (eds) *Le impronte di Robinson. Mass media, cultura popolare, educazione*, Turin: Elledici.

Cesareo, G. and Rodi, P. (1996) *Il mercato dei sogni*, Trento: Mondadori.

Giannatelli, R. (ed.) (1996) *Chiesa e comunicazione sociale. I documenti fondamentali*, Turin: Ellenici.

Gonnet, J. (1995) *De l'actualité à l'école. Pour des ateliers de démocratie*, Paris: Colin.

Greco, G. (2000a) *Mediamorfosi. Conversazioni su comunicazione e società*, Rubbettino: Soveria Mannelli.

——(2000b) *Socializzazione virtuale. Bambini e tv nei nuovi scenari tecnologici*, Rubbettino: Soveria Mannelli.

——(2004) *L'avvento della società mediale. Riflessioni su politica, sport, educazione*, Milan: Franco Angeli.

Hall, S. and Whannel, P. (1967) *The Popular Arts*, Boston: the Beacon Press.

Hart, A. (1998) *Teaching the Media: International Perspectives*, Mahwah, NJ: Erlbaum.

Masterman, L. (1985) *Teaching the Media*, London: Routledge.

——(1997) *A scuola di media. Educazione, media e democrazia*, La Scuola: Brescia.

Mattelart, A. (1998) *La comunicazione globale*, Rome: Edizione Riuniti.

Merton, R.K. (1957) *Social Theory and Social Structure*, Glencoe, IL: Free Press.

Morcellini, M. (1997) *Passaggio al futuro. Formazione e socializzazione tra vecchi e nuovi media*, Milan: Franco Angeli.

——(ed.) (2000) *Il mediaevo: TV e industria culturale del Ventesimo secolo*, Rome: Carocci.

Pavlic, B. (1987) 'UNESCO and Media Education', *Educational Media International*, 24: 32.

Piromallo Gambardella, A. (1997), *Costruzione ed appropriazione del sapere nei nuovi scenari tecnologici*, Naples: CUEN.

Rivoltella, P.C. (2001) *Media Education. Modelli, esperienze, profilo disciplinare*, Rome: Carocci.

Rivoltella, P.C. and Ardizzone, P. (eds) (2007) 'New Media Education', *Scuola e Didattica,* 15.

Rivoltella, P.C. and Marazzi, C. (2001) *Le professioni della Media Education*, Rome: Carocci.

Salzano, D. (ed.) (2000) *Comunicazione ed educazione. Incontro di due culture*, Naples: L'isola dei ragazzi.

Scurati, C. and Caimi, L.(1991) *Profili nell'educazione: ideali e modelli pedagogici nel pensiero contemporaneo*, Milan: Vita e Pensiero.

Thompson, J.B. (1998) *Mezzi di comunicazione e modernità*, Bologne: Il Mulino.

Part II

Scholarly settings

Chapter 4

ICT and relationships

Promoting positive peer interactions

Antonella Brighi, Manuela Fabbri, Luigi Guerra and Elena Pacetti

Introduction

This chapter aims to evaluate the potential of existing techniques that, before being adopted, should be selected and eventually integrated or modified to suit the different educational situations. For this purpose, we will provide a summary of good practices concerning structural innovation in education, as experienced in Italian schools: school–community relations, the promotion of relations with families, the stimulation of meta-cognitive learning, school twinning, interdisciplinary learning, and management of assessment and remedial activities. Even if apparently these latter topics may pertain mostly to cognitive processes, these ideas share a socio-constructivist approach, which underlines the strict relationship between learning and collaborative activity, hence focusing on relationships between the knower and the known, as stated by Vygotsky (1978) and Jerome Bruner (1987). This states that knowledge is generated through social interaction, and through this interaction individuals gradually progress through their levels of knowing. Social constructivism is currently the most accepted epistemological position associated with computer-based learning (Kanuka and Anderson, 1998), given the crucial role of relationships in virtual communities, and in this chapter many examples of how this theoretical framework can be implemented in school activities will be given.

New ICTs, both in their online and offline dimensions, represent a potential tool for an extraordinary qualification of the educational experience, framing the most recent development of our knowledge society; Koschmann (1996), discussing new technologies and education, argues:

> Traditional theories of learning treat learning as a concealed and inferred process, something that takes place inside the learner and only inside the learner. CSCL [Computer-Supported Collaborative Learning] research has the advantage of studying learning in settings in which learning is observably and accountably embedded in collaborative activity.
>
> (Koschmann, 1996, p. 19)

To successfully implement and use CSCL in natural settings, however, one has to resolve technical, organizational, and pedagogical challenges. This is possible only if technologies are placed at the service of critical models of teaching mediation, in other words, if they are located within an analysis which takes into account the interaction between the technical and psycho-pedagogical dimensions. On the one hand, this analysis may allow us to evaluate the potential of existing techniques that should be selected and eventually integrated or modified to suit the different educational situations; on the other hand, if necessary, this analysis could promote the invention of new techniques or ways of using existing techniques and technologies, thereby reversing the process that now seems to go inexorably and uncritically in a 'technocentric direction' from the development of techniques and technologies to their use in pedagogy. Thus, instead of focusing extensively only on technology, one should turn towards the social settings that support the implementation and use of technology, since creating new learning environments or learning communities, in fact, is not just a matter of implementing and putting into use new technology but in many cases, also applying simultaneously new practices of learning and instruction. Collaborative virtual learning environments rely on the assumption that knowledge is a complex entity that is shaped by social context, not a simple product to be transmitted or shared. Knowledge is the combination of information, context, and experience. According to this approach, some authors (e.g. Lipponen, 2002; Daniel *et al.*, 2003) acknowledge the founding role of relationships in learning in virtual spaces, and underline how the dimensions of social network structures, trust, common norms and cooperation between members may lead to the construction of a shared 'social capital', or to the idea of 'togetherness': 'Effective interactions among these factors can help group members solve collective problems, collaborate, learn and develop social relationships that can tie them to a community' (Lipponen, 2002, p. 74). This implies that knowledge is generated through social interaction, and through this interaction individuals gradually progress in their levels of knowing. This view of learning as part of relational experiences is more prominent in socio-cultural learning theories and constructivist theories of learning (Vygotsky, 1978). It is not the case that Vygotsky's approach, with its emphasis on the influence of cultural and social contexts in learning, and the theories inspired by social constructivism, are currently the most accepted epistemological positions associated with computer-based learning (Kanuka and Anderson, 1998).

Socio-constructivist learning

The school should promote a socio-constructivist learning environment and Web 2.0 is proving to be a powerful tool for collaboration and the social construction of knowledge. Some examples of the use of new ICTs for this purpose are: the design and social use of a class website in which students and teachers are both content users and producers; opening up communication with other classes and other schools, with families; creating a virtual space so that it becomes the virtual square of the classroom where they can meet, talk and exchange useful information, and share experiences. Another example: the design and management of educational applications such as forums, wikis, blogs to organize instructional content (text, images, video), in order to be able to represent and express their ideas in a critical perspective, or to reorganize contents produced by others according to the needs of the class itself; moreover: the construction of shared virtual spaces for collaborative learning activities in order to solicit active participation and collaboration that is respectful of the individual. In this way, everyone feels involved in the task and responsible for the product. In other words, the advanced network applications designed for educational purposes may support users' cognitive activities by providing advanced socio-cognitive scaffolding by pre-structuring the kinds of contributions learners can make, and by including community-building tools (Pea *et al.*, 1999). ICTs allow us to organize the knowledge and the skills in a collaborative and decentralized way, as they are based on sharing information, resources and hyper-links between individuals who cooperate among peers with no hierarchical structures. ICTs create a true peer education through which students may learn to negotiate roles and methodologies along the knowledge-building process. Learning with ICTs can bring together the diverse viewpoints in the group and teach students to work together, not only to reach a common view, but also to learn how to take into account different representations of knowledge and its complexity (Pacetti, 2010); therefore, it means being able to mediate among different opinions, to develop projects that integrate the different ways of thinking and behaving, and to consider diversity as a value added to the work of the group (Granieri, 2005).

The relationship between school and the local social, cultural and economic environment

A school that operates in the context of an integrated training system (Frabboni *et al.*, 1999), a system, then, which recognizes and affords equal dignity to the extra-scholastic educative agencies in its area, is a school that is able to relate, communicate, collaborate and integrate itself with the educational opportunities of its external environment, assuming a relationship of complementarity. The educational resources of the virtual space may be either intentionally educational and institutional (museums, libraries, recreation centres, sports centres, cultural and recreational offerings, in addition to other connections) or unintentionally educational and non-institutional in nature (the opportunities for viewing landscape or monuments, for shopping or network-related services). The school must recognize and value in its curriculum the importance of experiences carried out by students out of school (their expertise, knowledge, motivations) and must also 'network' with educational, social, cultural, environmental opportunities offered by the area where the school is situated, creating strong alliances with the various educational agencies in order to ensure the quality and continuity of the experience with school knowledge. ICTs, as tools that enhance the connection and interaction, allow the implementation of a network between the school and the extra-scholastic system, creating an educational hypermedia city where people can also meet virtually and share experiences. Some examples include: using the internet to gather information about cultural heritage, or the use of the network in an interdisciplinary way (history, geography, literature, arts, foreign language ...) to search for quality information, to learn before embarking on a study tour or to learn about the existing cultural projects, or to plan a route between museums, monuments, environmental resources. Another example may be the use of an interactive whiteboard as a hypertext container of the city, or to build a multimedia archive of the opportunities offered by the city from the personal experiences of the pupils. Moreover, a blog can be a collective volume on the city, or a tool to systematize and document the cultural, environmental and social policies implemented by the school, the students and their families, extended to all citizens in the interests of communication, sharing and expansion of knowledge.

The relationship with the families

The relationship between the school and the students' family (or between the school and the parental system since the concept of family is undergoing a rapid evolution) is one of the crucial points of the contemporary school, dealing with the opposing needs of shared pedagogical models and the necessity to acknowledge each family's culture and history, in other words, its diversity. These features are due not only to migration processes, but also to post-modernity and its effects on the habits, behaviour and values of the 'historical' residents. As a consequence, some Italian laws which were passed around 1980 in order to grant parents representative participation in the decision-making processes in schools appear to be completely inadequate: the dominant socio-cultural climate and some aspects of recent reforms have led many parents to consider the school as a kind of service 'on individual demand', on the one hand, and, on the other, the school personnel tend to consider the relationship with families as a mere exchange of information. Therefore, new forms of parents' involvement in the educational choices operated by the schools must be found. In this framework, ICTs, in both their online and offline dimensions, can be an essential support when organizing an effective and efficient exchange of information between the school and the family. At the same time, they can support systematic sessions of 'parent training' by building an environment where meaningful ongoing debate on educational issues can be carried out. ICTs can also be used to manage the inevitable micro/macro conflicts between schools and families and among family members, and to accompany the necessary form of solidarity for family problems, ranging from the afternoon homework arrangements to the organization of school trips or other events, to improve the socialization process outside school. Finally, ICTs may eventually be involved in creating and sharing experiences through the publication of the virtual school newspaper, reporting abilities in a systematic way about experiences at school (Caprara and Guerra, 2004).

Metacognitive learning stimulation

The educational perspective of metacognition broadly interprets education as a process of building the skills of the students and the group, with particular attention to the process, in other words, metacognition is conceived of as the acquisition by the individual of instruments of cultural elaboration directly reusable in different contexts or as the ability of 'learning to learn' (Cornoldi, 1995). Nowadays, with the increasingly vast amounts of information, ICTs may imply the risk of lack of attention to metacognitive skills, which have been replaced and buried under an 'avalanche' of new notions. On the other hand, without underestimating the risks reported, a technologically aware model can introduce computers into schools to help pupils make the most of their metacognitive values. Values that ICTs have in large measure, are as a means of mediation between the individual and knowledge, as ways of working if they rely on rules and which become more effective to the extent that individuals manage these working rules. Computers can support a metacognitive approach to teaching, since they can lead the students 'to learn to learn' through the acquisition of the ability to interpret, organize and structure information, to become increasingly aware of the processes underlying these operations, so they can manage these skills increasingly autonomously (Varisco and Grion, 2000). The technological problem is how precisely to plan and conduct this kind of didactics, avoiding a merely reproductive attitude in the student, influenced by the user-friendly possibilities of ICTs that may result in a mindless and 'acritical' use of them, as if the machines were working on their own:

> Making thinking visible by writing and sharing should help students to reflect on their own and others' ideas and share their expertise. Shared discourse spaces and distributed interaction can offer multiple perspectives and zones of proximal development (ZPD) for students with varying knowledge and competencies. CSCL environments can also offer greater opportunities to share and solicit knowledge. Furthermore, the database can function as a collective memory for a learning community,

> storing the history of knowledge construction processes for revisions and future use.
>
> (Koschmann, 1996, p. 22)

Therefore, according to this viewpoint, one can consider the possibilities offered by new technologies to design and document the experience hypertextually, to socialize it critically with the interactive whiteboard, to explore it before living it or when evaluating it, through the use of hypertextual maps.

Twinning schools

Within the confines of the global knowledge society, it is appropriate that the school promotes advanced forms of twinning between schools and remote collaboration through the sharing and communication tools offered by the net. The encounter with another person who is different from me, in a virtual context, which, however, might also become real (twinning may provide opportunities to meet face to face), means promoting relations between different cultures. Relationships do not cancel the differences, but enhance them. In the tension between opposing and present demands of globalization and localization, ICTs can become a meeting place where we can know each other and 'recognize' one another, enhancing our own individual and collective identity, dealing with realities which are very different from those normally experienced, teaching us to listen and to share (information, ideas, materials, cultures), creating forms of critical cross-breeding of cultures which provide skills to live as active and responsible citizens in contemporary society. Some possible examples of the use of ICTs in this domain can be the realization of a class website to exchange information about history, traditions, cultures, areas; the use of different telematic instruments in synchronous and asynchronous ways (chat, forums, video conferencing, mailing lists, blogs, wikis), to learn to collaborate and communicate while respecting the other, and sharing skills and knowledge with other users; participation in websites of national and international exchanges; becoming a community of learners keeping one's own singularity and uniqueness, and discovering and respecting the other. Critical use of the net teaches us to discriminate when accessing information, to be able to assess and compare the different contents of a webpage, to know how to use new media with a critical and careful eye, and to be able to write and produce online contents.

Support for an 'inclusive school'

The inclusive school is one that can meet the challenge of embracing old and new diversities, from those related to individual intellectual or physical characteristics to those associated with social disadvantage, or one that is structurally related to the myriad forms of difference (gender, language, culture, customs). Designing and implementing inclusion means realizing an educational environment that can respond to the dialectically opposing rights of equality and diversity, designing teaching methods to allow all learners to acquire the skills and shared ownership of knowledge essential for their integration into society, and at the same time being open to educational experiences that may feature specific aims, resources, and cultural views of the individual learner. In this context, ICT seems to be particularly effective in designing an educational environment which is able simultaneously to provide individualization and therefore structured learning paths, moments of formative assessment, time and materials that will allow everyone to achieve the same goals, as well as being open to 'customization', capable of allowing individual learners or groups of learners to achieve personal educational experiences, based on their own specific choices. The use of computers in the classroom can enable the management of teaching for 'heterogeneous classes' permanently ready to welcome diversity, involving learners and their families in increasing the value of individual differences and at the same time, in pursuing shared educational goals. Throughout the world, the concept of the inclusive school is used to refer also to a school that can welcome students who have specific dimensions of diversity, related to issues of psychic or physical disability. Regarding these learners, new technologies allow schools to reduce problems arising from the differing abilities of individuals through the proposal of special hardware or the use of specific software. The most interesting part of the use of hardware and software for the reduction of handicap made available by ICT is to see inclusion as a set of interventions aimed at the disabled student but necessarily involving the whole class, therefore overcoming the element of exclusive individual care that would give rise inevitably to the phenomenon of exclusion (for a review, see Cullen *et al.*, 2009; HaccaParlante, 2010; Tosello, 2010a, 2010b).

Interdisciplinary learning

The organization of knowledge in disciplines is one of the characteristics of the traditional school. There is no shortage in the history of education of those contributions (e.g. Bruner, 1966) which stress the importance of a disciplinary approach to knowledge, emphasizing the role of the specific disciplines as many 'open windows' on cultural heritage. In the literature there is also no lack of criticism of a disciplinary approach that fails to teach the complexity of particular cultural issues, as they are always meeting places between different disciplinary approaches. Certainly, the contemporary school, while remaining structurally discipline-oriented, cannot abandon the production of significant holistic learning experiences in which the contents of learning are addressed in a multidisciplinary and interdisciplinary way. In this framework, new technologies, through software, such as conceptual maps, hypertexts, or shared use of blogs and wikis, can permit the design of multidisciplinary education programmes, in which the same subject is addressed by different disciplines. Through the same tools, one can then support the design of interdisciplinary education programmes, where the same topic is treated in a partnership among several disciplines that combine their language and their research tools.

Finally, in more advanced didactic methodologies, ICT can support the development of assessment models able to measure and appreciate the achievement by learners of holistic skills in problem-solving projects and case studies.

Management of assessment and remedial work

Evaluation interprets the tasks of selection commonly attributed to the traditional school. The contemporary school, abdicating the former 'constitutional' task, should perform more 'promotion' of every learner instead of being only the provider of certification of differences, which inevitably are influenced by class differences. In this context, for 30 years, educational innovation has emphasized mainly formative assessment: that is, its need to be an instrument for the measurement of problems and the attempt by the school to resolve them through appropriate actions based on individualized tasks, from remedial courses to individual support. The use of ICTs in formative assessment in this context also directs the the work of the teacher who is able to critically adapt his/her teaching methods to the training needs of the individual learner. In this framework, ICTs allow extremely effective and efficient organization that will enable significant moments of self-assessment, for example, giving access to individual questioning. ICTs allow teachers to prepare materials finalized to specific formative needs, and to realize an instructional design that will enable learners to have a clear picture of the development tasks assigned to them. Using a plan that incorporates the theme of relations between teachers, school and family, ICTs can also build up systematic ways of comparative evaluation among different class teachers, to deliver information about evaluation and provide guidance to families.

Conclusion

The considerations above refer to frameworks of potential use of new technologies in which the elements of complete innovation and positive use of ICTs are, however, constantly faced with the problem of their potential inappropriate uses and with possible drifts in the direction of their passive use, which may weaken the critical capacity of individuals, Therefore, the centrality of new technologies in the definition of the quality of cognitive and social life for individuals contains its own intrinsic problems, embedded in the capacity of individuals, groups and social contexts to integrate such new tools within the existing and political models, that should firmly remain in the hands of the individuals who use them. ICTs represent structurally new environments and new contexts, which are incompatible with the previous existing, cultural and political planning: this means to reaffirm the need to be aware that any new technology, even a miraculous are, may hide the risks of uncritical use and result in a threat to the freedom, critical thought and creativity of those who use them.

Bibliography

Anderson, T.D. and Garrison, D.R. (1998) 'Learning in a networked world: new roles and responsibilities', in C.C. Gibson (ed.) *Distance Learners in Higher Education: Institutional Responses for Quality Outcomes*, Madison, WI: Atwood Publishing, pp. 1–8.

Bielaczyc, K. (2001) 'Designing social infrastructure: the challenge of building computer-supported learning communities', in P. Dillenbourg, A. Eurelings, and K. Hakkarainen (eds) *European Perspectives on Computer-supported Collaborative Learning: The Proceedings of the First European Conference on Computer-Supported Collaborative Learning*, University of Maastricht, pp. 106–14.

Bruner, J.S. (1996) *Toward a theory of instruction*, Cambridge, Mass.: Harvard University Press.

—— (1987) *Making Sense: The child's construction of the world*, London: Methuen.

Bruns, A. (2008) *Blogs, Second Life, Wikipedia, and Beyond: From Production to Produsage*, New York: Peter Lang Publishing.

Caprara, B. and Guerra, L. (2004) *Il giornale virtuale. Esperimenti in rete per la comunicazione interscolastica*, Bergamo: Edizione Junior.

Cornoldi, C. (1995) *Metacognizione e apprendimento*, Bologne: Il Mulino.

Cullen, J., Cullen, C., Hayward, D. and Maes, V. (2009) *Good Practices for Learning 2.0: Promoting Inclusion, an In-depth Study of Eight Learning 2.0 Cases, IPTS, European Commission*, available at: http://ftp.jrc.es/EURdoc/JRC53578_TN.pdf.

Daniel, B., Schwier, R. and McCalla, G. (2003) 'Social capital in virtual learning communities and distributed communities of practice', *Canadian Journal of Learning and Technology*, 29(3): 113–39.

Fabbri, M. (2005) *Empowerment e nuove tecnologie. Telematica e problematiche della devianza e delle dipendenze*, Bergamo: Edizione Junior.

Frabboni, F., Guerra, L. and Scurati, C. (1999) *Pedagogia. Realtà e prospettive dell'educazione*, Milan: Mondadori.

Garrison, R. and Anderson, T. (2003) *E-learning in the 21st Century: A Framework for Research and Practice*, New York: RoutledgeFalmer.

Granieri, G. (2005) *Blog Generation*, Bari: Editori Laterza.

Guerra, L. (ed.) (2010) *Tecnologie dell'educazione e innovazione didattica*, Bergamo: Edizioni Junior.

HaccaParlante (2010) *Disabili e web 2.0. Relazioni sociali e servizi, informarsi e informare, accessibilità*, Trento: Erickson.

Jenkins, H. (2006) *Fans, Bloggers, and Gamers: Exploring Participatory Culture*, New York: New York University Press.

Kanuka, H. and Anderson, T. (1998) 'Online social interchange, discord, and knowledge construction', *Journal of Distance Education*, available at: http://cade.athabascau.ca/vol13.1/kanuka.html.

Koschmann, T. (1996) *CSCL: Theory and Practice of an Emerging Paradigm*. Mahwah, NJ: Lawrence Erlbaum Associates.

Lipponen, L. (2002) 'Exploring Foundations for Computer-Supported Collaborative Learning', in *CSCL '02 Proceedings of the Conference on Computer Support for Collaborative Learning: Foundations for a Community*, pp. 72–82.

Nardone, R. (2007) *I nuovi scenari educ@tivi del videogioco*, Bergamo: Edizioni Junior.

Pacetti, E. (2010) 'Social software, Community e reti sociali. Vivere il Web 2.0', in L. Guerra (ed.) *Tecnologie dell'educazione e innovazione didattica*, Bergamo: Edizioni Junior.

Pea, R., Tinker, R., Linn, M., Means, J., Brandsford, J., Roschelle, J., Hsi, S., Brophy, S. and Songer, N. (1999) 'Toward a learning technologies knowledge network', *Educational Technology Research and Development*, 4: 19–38.

Tosello, D. (2010a) *Disabili 2.0 e lavoro: si può fare?* Available at: http://www.disabili.com/blog/tecnologie-assistive/disabili-2-0-e-lavoro-si-puo-fare/.

——(2010b) *Gli ausili invisibili dell'ISTAT. Perché in Italia non si investe sulle tecnologie per la disabilità?* Available at: http://www.disabili.com/blog/tecnologie-assistive/gli-ausili-invisibili-dell%E2%80%99istat-perche-in-italia-non-si-investe-sulle-tecnologie-per-la-disabilita/.

Varisco, B.M. and Grion, V. (2000) *Apprendimento e tecnologie nella scuola di base*, Turin: UTET.

Vygotsky, L.S. (1978) *Mind in Society: The Development of Higher Psychological Processes*, Cambridge, MA: Harvard University Press.

Zanetti, F. (2002) *Telematica ed intercultura: Le differenze culturali nelle contraddizioni del villaggio globale*, Bergamo: Edizione Junior.

Chapter 5

The use of interpersonal communication technologies to establish and maintain peer relationships

Maili Pörhölä and Henna Lahti

Introduction

Today, both children and adolescents can establish and maintain peer relationships in real (physical) and virtual spaces. The majority of children and adolescents in Western societies have daily access to at least one of the various applications of information and communication technology (ICT). Via mobile phones and the internet, they are able to communicate with their friends and unknown peers at any time of the day or night. Indeed, social interaction with peers appears to be the most important function of ICT for children and adolescents (Kaare *et al.*, 2007; Pelastakaa Lapset, 2009; Subrahmanyam *et al.*, 2002), suggesting that a substantial part of their cognitive and emotional experiences, learning, seeking play and having fun, and relationship and identity building takes place online, in virtual rooms.

Kaare *et al.* (2007) note that children in Norway aged 10–12 years use ICTs in their peer relationships for several constructive purposes including: (1) mediating practical information and messages; (2) mediating interests and hobbies; (3) seeking to establish contact with others; (4) mediating either or both strong symbolic and emotional contents (e.g. pictures as symbolic gift-giving, ritualized greetings, 'I like you' messages); (5) communicating with or about the opposite sex; but unfortunately, also for (6) bullying and harassing their peers. However, although the rapid development of ICT has provided new tools for causing harm to peers, for example, cyberbullying (e.g. Beran and Li, 2007; Juvonen and Gross, 2008; Smith *et al.*, 2008; Pure, 2009; Livingstone *et al.*, 2011), a great number of children and adolescents frequently use this technology to promote positive peer interaction and successful social networking.

How do children and adolescents use ICT in their daily life for peer networking and facilitating their peer relationships? Based on our examination of approximately 150 research reports, articles, and book chapters, we outline in this chapter the main findings of research conducted on the positive uses of ICT in children's and adolescents' peer relationships. This corpus of this research represents a variety of disciplines including communication studies, psychology, sociology, social psychology and education. Because of the space limits, we selected a reasonable number of literatures to be reviewed in this chapter. The chapter focuses on: (1) the ways in which children and adolescents use various applications of communication technology (e.g. instant messaging, chat rooms, online games, internet galleries, mobile telephones and e-mail) to establish and maintain peer relationships; (2) gender-specific differences in usage; and (3) the internet communities which they join. The chapter presents two applications to illustrate the kinds of internet-based social forums that have been developed in Finland for children and adolescents to promote their peer networking in positive ways and to provide support for their peer relationships.

Maintenance of peer relationships

As several studies indicate, the most important function of online communication for children and adolescents is maintaining and improving their existing peer relationships. Kaare *et al.* (2007) found that children in Norway, aged 10–12 years, mostly used mobile phones, and particularly text messaging, as a less expensive tool than phone calls, to keep in contact and share practical everyday information with their friends. However, they also used instant messaging, e-mail, and chat rooms for interaction with their friends, unknown peers and family members.

Kaare *et al.* (2007) conclude that children, once mobile phones became common tools for them, seem to have adopted a communication culture in which they feel they should be available for their peers all the time. They keep their mobiles constantly to hand, and respond immediately to any messages they receive. Mobile phones also seem to provide youth with a social meaning at insignificant times, such as travelling between home and a bus stop, as they can spend this time communicating with their friends (Caron and Caronia, 2007). Furthermore, as Kasesniemi and Rautiainen (2001) note, a friend group can, by means of their online interaction practices, create a unique communication culture, which outsiders cannot easily access.

When maintaining peer relationships, online communication appears to be an important tool for adolescents and teens as the internet enables contact to be made with both nearby and long-distance peers (Subrahmanyam *et al.*, 2002). Blais *et al.* (2008) found that instant messaging is positively associated with most aspects of relationship quality in adolescents' best friendships and romantic relationships. Similarly, Valkenburg and Peter (2007) found that adolescents who often communicated online with their known peers felt closer to their friends compared to those who communicated online less often with their friends. Furthermore, in a national survey in the United States, Lenhart *et al.* (2001) found that approximately half of the adolescent respondents felt the internet had improved their friendships. American adolescents also reported the internet had enabled them to keep in contact with long-distance peers. However, these adolescents did state the internet was not the best channel for establishing close friendships, unless the partner was already known.

Girls tend to use ICT for social interaction, sharing concerns and emotions, giving and asking for social support (McKay *et al.*, 2005; Oksanen and Näre, 2006; Joensuu, 2007; Stern, 2007), and to build their identity by uploading photos and different kinds of texts in web galleries (Joensuu, 2007). Boys use the internet for social interaction in playing online games and sharing files, e.g. music, movies, software (Joensuu, 2007; Livingstone *et al.*, 2011) and their ICT messages are characterized by informativeness and brevity of expressions (Kasesniemi and Rautiainen, 2001; Livingstone and Bovill, 2001). Blais *et al.* (2008) argue the internet probably adds to girls' already higher levels of relationship quality and enables them to further expand their socialization. Girls, compared to boys, would be more likely to benefit from the positive effects that internet communication has on their relationships.

Establishment of new peer relationships

Children and adolescents also use ICTs to establish new peer relationships. Websites that focus on various hobbies and fan sites of music bands provide a wide variety of virtual rooms for meeting new peers with similar interests (Sjöberg, 2002). While boys are more engaged in sharing experiences of computer games and visiting the fan sites of well-known films, girls show more interest in specific pop stars and films, or visit specific pet sites (Kaare *et al.*, 2007).

Every second adolescent has used some form of ICT (e.g. instant messaging, e-mail, or chat rooms) for interaction with a person they have never met face-to-face (Lenhart *et al.*, 2001; Wolak *et al.*, 2002). Wolak *et al.* studied adolescents aged 10–17 years, 25 per cent of whom reported having casual online friendships, and 17 per cent as having formed at least one close online relationship (a close relationship, romantic relationship, or a relationship leading to a face-to-face meeting) during the past year. A further 5 per cent claimed to have established more than one type of close online relationship in the past year. An online relationship most typically starts in a chat room or in an online game or role play and interaction is then transformed into a more interpersonal level by continuing communication via instant messaging, e-mail, text-messaging, or phone calls (Caron and Caronia, 2007; Joensuu, 2007; Wolak *et al.*, 2002). More than two-thirds of adolescents' online relationships are cross-gender relationships (Wolak *et al.*, 2002).

Sometimes the partners of an online relationship eventually end up in a face-to-face meeting. A wide survey conducted in 25 European countries reveals that 9 per cent of adolescents had met their internet partners face-to-face (Livingstone *et al.*, 2011). Wolak *et al.* (2002) found that adolescent internet partners in the United States who had actually met were mostly living in close proximity and their parents knew about the new relationship. Face-to-face rendezvous were mostly held in public places and the new peer was met in the presence of a friend. Hence, the relationships which start on the internet seem to develop in a safe manner for most adolescents.

However, the internet also provides children and adolescents with an arena for identity building, concealing their real identity and even allowing them to adopt a false identity. Participants in a chat room or in an online game can, for example, state their age and gender without revealing their identity; they can give false information about some of their characteristics such as their age or home town; or they can pretend to be someone else (Smith and Curtin, 2001; Sjöberg, 2002; Kaare *et al.*, 2007). Lenhart *et al.* (2001) found that one-third of teenagers have sometimes pretended to be someone else in their e-mails or instant messages. The internet also enables users to search for a sexual identity, as anonymity provides the opportunity to pretend to be a representative of the opposite sex (Kaare *et al.*, 2007).

Adolescents also use ICTs for different purposes in their romantic relationships. For example, children aged 10–12 years use text messaging by mobile phone to exchange love messages (Kaare *et al.*, 2007). Lenhart *et al.* (2001) found that 17 per cent of the adolescents used instant messaging when starting a dating relationship by, for example, asking the person for a date. Furthermore, these adolescents felt that the confession of love was easier to make online than face-to-face. On the other hand, 13 per cent reported having used the same tool to break up with their romantic partner. A text message to break up the relationship can also be sent by mobile phone (Kasesniemi and Rautiainen, 2001). Hence, ICT is used by youth both to establish and end peer relationships.

Applications of virtual rooms for children and adolescents

Habbo Hotel

Habbo Hotel, the world's largest virtual world for teenagers, is based on social interaction and meeting other people online via their avatars (Habbos). The Finnish Sulake Corporation describes this product on their web pages:

> Habbo Hotel is a place to meet new and existing friends, play games and simply have fun. It is a richly colorful, multi-dimensional virtual community and game environment which users join by creating a fully customized online character called a Habbo. From there, Habbos can explore many public spaces in the virtual environment, play a variety of games, train pets, connect with friends, decorate their own rooms and Habbo homepages, and have fun through creativity and self-expression.

> The Habbo experience consists of many parts, which are all inextricably linked.
>
> (Sulake Corporation Oy, n.d.)

The core of the experience is the virtual world called Habbo Hotel. Millions of teenagers around the world meet each other every day in Habbo's cafés, clubs, at the pool area or in the millions of private rooms created by the Habbos themselves. Entering the Habbo world is free, but to access premium services, users need to buy Habbo credits, which is the virtual world's currency. Credits are used to pay, for example, for virtual furniture and elements of the user's homepage. In addition to the actual virtual world, all community members have their own Habbo home pages where they can present their virtual life. These pages are highly customizable and can be used, for example, to share information or form different kinds of groups (ibid.).

Initially designed in Finland in 2000 as a hobby project by Sampo Karjalainen and Aapo Kyrölä, by 2010, Habbo Hotel had developed into a global application with local communities in 31 countries and in 11 different languages. Habbo has attracted over 168 million registered users around the world, of whom 65 per cent are aged 13–16 years and 65 per cent visit the Habbo Hotel daily with an average visit lasting for 43 minutes (ibid.).

Habbo Hotel is really a whole world with its own language and slang and ways of behaving, so it takes a while for users to become integrated. Habbo Hotel is an option for a normal chat because instead of writing comments to anonymous nicknames, a user in Habbo Hotel can see the other Habbos all the time. The avatars (Habbos) always reveal something about the user.

Establishing relationships in Habbo Hotel starts by asking other Habbos to be your friends. Accepted friends can then be seen in the users' profile as a list, so it is easy to contact them later or, for example, follow them into a room where they are staying, or start talking to them in a private messenger window. Habbos also invite their friends to their own rooms, which they may have furnished and decorated nicely.

Young people in Habbo Hotel often discuss their hobbies, school, and any kinds of everyday subjects with their virtual friends. Sometimes they can just try to gain others' attention by, for example, writing a pointless series of numbers or letters. In addition to chatting with other children and adolescents, Habbo Hotel provides a way just to spend time. Indeed, just as in the real world, discussions are sometimes pointless and do not lead to any reasonable outcomes.

Chat rooms in Habbo Hotel are no different from any other chat rooms in that it is difficult to know when the users are telling the truth about themselves and chatting with a serious attitude. An example of a confusing discussion which we witnessed occurred in a public room where a girl and a boy talked about suicide and how and why they both had tried to do it. In the middle of the conversation, one of the avatars suddenly disappeared leaving the other alone. After a while the avatar returned and started talking about a totally different topic, no longer answering any of the questions posed by the other participant about the former topic. It seemed obvious that the person behind the avatar who had left and returned to the room had only been having fun, whereas the other participant had been revealing personal issues and was oriented to a serious discussion. This example emphasizes the need to talk with children about how they should act and communicate online, and how they can protect themselves from being hurt by their virtual peers.

IRC-Galleria

Sulake Corporation Oy also own IRC-Galleria, which is the most active social networking service in Finland. Users in IRC-Galleria choose a nickname for themselves, create their own profile and upload pictures. Anyone online can see the uploaded pictures and registered members can comment on others' pictures, which creates an interactional aspect to the service. In addition to commenting on photos, users can publish longer notes and blog texts in their profiles. Comments posted to others' pictures or profiles can be related to photos, or just any comments concerning the users' lives, interests and hobbies. Any time someone posts a comment to another user's profile, the sender's nickname appears and by clicking the name, the user can move to the sender's profile. Using the service is free, but members can buy special effects to add to their photos or VIP services that enable some additional functions (Sulake Corporation Oy, n.d.).

There were approximately 658,000 young people aged 15–24 years living in Finland at the end of 2008 (Väestörekisterikeskus, 2009), of whom almost 76 per cent used IRC-Galleria weekly, and 60 per cent of these users were older than 18 years. The 500,000 young people registered with IRC-Galleria write 1.4 million comment messages, upload 75,000 photos and write 50,000 blog entries daily. On average, registered individuals spend approximately 30 minutes a day using IRC-Galleria. The minimum age limit for users is 12 years and children under that age cannot be registered even with the permission of their parents (Sulake Corporation Oy, n.d.).

According to Oksanen and Näre (2006), IRC-Galleria is particularly popular among young girls, who search for their identities online by uploading provocative pictures and maybe also publishing blog texts. These young girls expose themselves to danger through abusive comments and critiques. Using data from 13 semi-structured interviews and participant observations of IRC-Galleria users, Lehtinen (2007) examined the ways in which young people maintain and extend their social networks on IRC-Galleria. She concluded that the pictures uploaded to users' profiles present teenagers' personal interests and objects which they think are appreciated in their own subculture and social networks.

Observations that we conducted on IRC-Galleria indicated that interaction between individuals usually starts tentatively but eventually these interactions also involve becoming acquainted with the friends of this new friend. Getting to know new people usually started with writing a neutral comment like 'nice photo!' or by asking 'how are you?' After a few comments, the participants might continue their interaction by instant messaging. The topics that youth on IRC-Galleria talk about vary widely. An initial comment may applaud someone's new look and the next comment can be an invitation to a party. Sometimes chatting is just filling in time and then there is no specific or special content in the discussion. Most of the comments seem to be from people users already know. Most are simply one sentence comments 'Oh, you looked really nice today–as always!', or related to former comments and discussions of events that happened in school on an earlier occasion.

Youth work in Habbo Hotel and IRC-Galleria

Recently there has been a trend in online youth work in Finland and many non-governmental organizations (NGOs) have extended their work from authentic real-life fields to online environments. One example is Netari.fi, which focuses on contacting young people who spend a lot of time online. Depending on the time, date and forum, policemen, nurses and youth workers are available for real-time online chats with adolescents to answer their questions. Netari.fi initially served only the Helsinki area but today there are professionals from 24 different municipalities around Finland.

Netari.fi organizes, several times a week, real-time online discussions and chats on both Habbo Hotel and IRC-Galleria, and a themed chat once a week on IRC-Galleria, on issues interesting to youth. Once a week, there is also a chat, in which policemen answer the questions posed by the young people; these chats are primarily directed at young people aged between 15 and 18 years. One specialty of the youth work in Habbo Hotel is Hubu, which is a virtual bus where teenagers can discuss drugs and alcohol with a trained adult. Hubu is organized by an NGO, and is part of the preventive drug work among Finnish teenagers (Elämä on parasta huumetta ry, 2010).

There is a definite need for youth work services online. According to the web pages of Netari.fi, the data for 2009 show approximately 159,000 young people used Netari services. Besides the themed chats, Netari's online youth facilities include special rooms (spaces) where, in the presence of professional youth workers, young people can hang around with others and talk (Helsingin ja Oulun kaupungin nuorisoasiainkeskukset, n.d.).

Results of online gallery use

The Mannerheim League for Child Welfare conducted a broad online survey examining the use of Finnish online galleries by young people in 2006 and 2007 (see Mustonen and Peura, 2007). Two-thirds of the respondents were girls, which corresponds to the data of users who log into gallery services.

The results showed that galleries have become an important forum to maintain and strengthen young people's existing peer relations. Almost every respondent (91 per cent) reported having talked on gallery sites with people they knew. Furthermore, almost one in two had found a new friend from gallery sites and one in ten had found a girlfriend or boyfriend through these gallery web pages.

There are, however, also clear negative effects of web galleries, such as taunting other users, sexually distracting comments, and manipulating other users' pictures without asking permission. One-fifth (19 per cent) of the respondents stated they had been bullied online during the past year. Cyberbullies were usually people whom the victim knew, for example, from school, and every fifth school bullying case continued in one online format or another after school (ibid.).

Although the use of galleries and other websites has increased, the negative effects of the services have not increased to the same extent. Education and talking about the possible risks inherent to the media, as well as learning by experience, are crucial steps toward young people's wisdom online. It is worth noting that young people are also in a key position to transfer knowledge about the internet and its applications to older generations. Mediaskooppi (Mediascope), through an informal web-based network for young people who are involved and interested in digital media cultures, aims to create a dynamic dialogue between young people and adults (Mediaskooppi, 2007).

In order to achieve this goal, the network arranges digital media education for schools and other institutions and helps adults to understand what kind of sites and applications adolescents use online and why. The activities of Mediascope consist of lectures in, for example, seminars, public libraries and schools. The idea of the network is to let those who know best talk about the sites and applications. Those who know best are young people who have gained the most up-to-date and diversified knowledge about the digital youth cultures emerging on the Internet. Mediascope was founded in 2007, following a high school web course on media criticism, and, although financed by the Ministry of Education in Finland, works independently, co-operating with educational organizations (ibid.).

Mediascope has been a model for many multi-generational media education projects. Hundreds of young people have been involved in its activities so far, and some 50 of them (mainly high school students) are actively involved in training; some of which have also been delivered via the internet (Jaakkola, pers. comm., e-mail, 26 April 2010).

Discussion

As indicated in this literature review, children and adolescents use multiple forms of information and communication technology to establish peer and dating relationships and to maintain and enlarge their social networks. They use this technology as a communication channel to enlarge their physical environments into virtual rooms where they can continue sharing their social experiences with known and unknown peers. Individuals regulate the cycle of real-life relationships (introduction, development and closure) in these virtual rooms. However, the anonymity provided by these virtual rooms also enables kinds of communication which would not be possible in a face-to-face environment.

While most children and adolescents use ICT in their peer relationships for constructive social purposes, recent research clearly indicates that some of them use the same tools for cyberbullying (e.g. Beran and Li, 2007; Juvonen and Gross, 2008; Smith *et al.*, 2008; Pure, 2009). Cyberbullying ensures a substantial number of youth suffer from the disadvantages of ICT in their peer relationships. We do not yet know whether or not individuals are engaged in constructive and destructive uses of ICT in their peer relationships or indeed whether some individuals have a tendency to be either constructive or destructive. What kind of ICT uses are characteristic of those who bully and those who are victims, either or both online and offline? More research is needed to answer these questions and to find ways in which these individuals could be encouraged and guided to engage in positive peer networking by means of ICT.

This chapter has discussed two websites, as examples of internet-based social forums, which have been developed to promote children's and adolescents' constructive peer networking. The aim of doing so was to shed some light on the possibilities that exist to support the positive development of children's and adolescents' peer relationships. Some of the ideas provided by these examples could also be applied to the development of specific sites for those who are engaged in either cyberbullying or are its victims. These social forum sites could help to disseminate information, for example, about the legal aspects and consequences of cyberbullying, and the suffering that cyberbullying causes the victims, and about effective coping methods and support available for victims. These types of websites could provide an interactive forum for children and adolescents to share their negative experiences of bullying, to gain peer support, as well as enlarge their understanding of their peer relationships in a safe virtual environment, which would be guarded by adults. Professionals of different kinds (e.g. youth social workers, communication experts, psychologists, and police and legal experts) could be on call at particular hours, discussing online with youth their concerns and giving them advice. The use of a common format with similar contents with country-specific languages would make this type of application cost-efficient.

References

Beran, T. and Li, Q. (2007) 'The relationship between cyberbullying and school bullying', *Journal of Student Wellbeing*, 1(2): 16–33.

Blais, J., Craig, W., Pepler, D. and Connolly, J. (2008) 'Adolescents online: the importance of Internet activity choices to salient relationships', *Journal of Youth and Adolescence*, 37(5): 522–36.

Caron, A.H. and Caronia, L. (2007) *Moving Cultures: Mobile Communication in Everyday Life*, Montreal: McGill-Queen's University Press.

Elämä on parasta huumetta ry (2010) *Hubula*. Available at: http://www.hubu.fi/ (accessed 21 May 2010).

Helsingin ja Oulun kaupungin nuorisoasiainkeskukset (n.d.) *Netari.fi*. Available at: www.netari.fi (accessed 10 May 2010).

Joensuu, M. (2007) *Nuorten Internet-yhteisöt ja Niiden Merkitys Nuorisotyön Kannalta*, Helsinki, Finland: Humanistinen ammattikorkeakoulu.

Juvonen, J. and Gross, E.F. (2008) 'Extending the school grounds? Bullying experiences in cyberspace', *Journal of School Health*, 78: 496–505.

Kaare, B.H., Brandtzæg, P.B., Heim, J. and Endestad, T. (2007) 'In the borderland between family orientation and peer culture: the use of communication technologies among Norwegian tweens', *New Media & Society*, 9(4): 603–24.

Kasesniemi, E. and Rautiainen, P. (2001) *Kännyssä Piilevät Sanomat: Nuoret, Väline ja Viesti*, Tampere, Finland: Tampere University Press.

Lehtinen, V. (2007) *Maintaining and Extending Social Networks in IRC-Galleria*. Available at: https://oa.doria.fi/handle/10024/7282 (accessed 21 May 2010).

Lenhart, A., Rainie, L. and Lewis, O. (2001) *Teenage Life Online*, Washington, DC: Pew Internet & American Life Project.

Livingstone, S. and Bovill, M. (2001) *Children and their Changing Media Environment: A European Comparative Study*, Mahwah, NJ: Lawrence Erlbaum.

Livingstone, S., Haddon, L., Görzig, A. and Ólafsson, K. (2011) *Risks and Safety on the Internet: The Perspective of European Children. Full Findings*, London: LSE, EU Kids Online.

McKay, S., Thurlow, C. and Zimmerman, H.T. (2005) 'Wired whizzes or techno-slaves? Young people and their emergent communication technologies', in A. Williams and C. Thurlow (eds) *Talking Adolescence: Perspectives on Communication in the Teenage Years*, New York: Peter Lang.

Mediaskooppi (2007) *Mediaskoopin Esittelydiat Englanniksi*. Available at: http://www.mediaskooppi.net/aineistot/mediascope.pdf (accessed 23 April 2010).

Mustonen, A. and Peura, J. (2007) *Netin Kuvagalleriat Nuorten Elämässä. Nettikyselyn Tuloksia 2007. Tulosten Vertailua 2006 ja 2007*. Available at: http://www.mll.fi/@Bin/11904188/Netin+kuvagalleriat+nuorten+el%c3%83%c2%a4m%c3%83%c2%a4ss%c3%83%c2%a4.pdf (accessed 23 April 2010).

Oksanen, A. and Näre, S. (2006) *Lapset Pelissä: Virtuaaliviidakon Ansat*, Jyväskylä, Finland: Minerva.

Pelastakaa Lapset (2009) *Lapsen Ääni 2009*, Helsinki, Finland: Pelastakaa Lapset ry.

Pure, R.A. (2009) 'Uncertainty and psychological consequences due to cyberbullying', in T.A. Kinney and M. Pörhölä (eds) *Anti- and Pro-Social Communication: Theories, Methods, and Applications*, New York: Peter Lang.

Sjöberg, U. (2002) 'Screen rites: a study of Swedish young people's use and meaning-making of screen-based media in everyday life', Lund, Sweden: University of Lund.

Smith, P.K., Mahdavi, J., Carvalho, M., Fisher, S., Russell, S. and Tippett, N. (2008) 'Cyberbullying: its nature and impact in secondary school pupils', *Journal of Child Psychology and Psychiatry*, 49: 376–85.

Smith, R. and Curtin, P. (2001) 'Children, computers and life online: education in a cyber-world', in I. Snyder (ed.) *Page to Screen: Taking Literacy into the Electronic Era*, repr. edn, London: Routledge.

Stern, S.T. (2007) *Instant Identity: Adolescent Girls and the World of Instant Messaging*, New York: Peter Lang.

Subrahmanyam, K., Greenfield, P.M., Kraut, R. and Gross, E. (2002) 'The impact of computer use on children's and adolescent's development', in S.L. Calvert, A.B. Jordan and R.R. Cocking (eds) *Children in the Digital Age: Influences of Electronic Media on Development*, Westport, CT: Praeger.

Sulake Corporation Oy (n.d.) *Sulake*. Available at: http://www.sulake.com/ (accessed 29 January 2010).

Väestörekisterikeskus (2009) *Taskutieto 2009*, Helsinki, Finland: Väestörekisterikeskus. Available at: http://www.vaestorekisterikeskus.fi/vrk/home.nsf/files/Taskutieto_2009/$file/Taskutieto_2009.pdf (accessed 10 May 2010).

Valkenburg, P.M. and Peter, J. (2007) 'Preadolescents' and adolescents' online communication and their closeness to friends', *Developmental Psychology*, 43(2): 267–77.

Wolak, J., Mitchell, K.J. and Finkelhor, D. (2002) 'Close online relationships in a national sample of adolescents', *Adolescence*, 37(147): 441–55.

Chapter 6

Prosocial use of the internet in adolescence

Rosario Del Rey, Virginia Sánchez and Rosario Ortega

Introduction

The new information and communication technologies have revolutionized the way we as humans relate to the world around us. However, this process, known as the technological revolution (Simone, 2000) requires an increasingly greater need for a fuller understanding of how populations, and in particular minors, access and employ ICTs (especially the internet) in order to avoid the risks of exposure to its potential dangers in terms of their social and psychological development (Livingstone, 2009).

Most reports and studies emphasize the potential risks in inappropriate use of ICTs. In the specific case of Spain, this has been corroborated by the latest Eurobarometer report (European Commission, 2008) which shows how Spanish parents are among the most preoccupied with the way that ICTs expose their chidren to inappropriate or unpleasant content or experiences such as cyberbullying.

Notwithstanding the above, the new technologies offer a wide variety of possibilities: in education and training, civic activity, creativity and identity, and in social networking (Livingstone and Haddon, 2009). They clearly offer minors new spaces for communication and leisure pursuits, and have made children important contributors to the digital literacy of the older members of their families. Great efforts are therefore being made to make access available to larger numbers of children and young people. In 2007, for example, the theme of the World Telecommunication and Information Society Day was 'connecting the young'.

Internet and mobile phone use by children, adolescents and young people

In Spain, research into how minors use ICTs is on a par with that being carried out in other European countries (Garmendia *et al.*, 2009). In general, Spanish minors appreciate the possibilities offered by ICTs, particularly the internet, where they can establish and cultivate friendships, and mobile phones, which offer liberty, intimacy, freedom from control, the chance to express themselves in an uninhibited manner, flexibility to meet and engage in activities with others, etc. (INJUVE, 2004). As in the rest of Europe, most research is descriptive, although some qualitative studies do exist which have attempted to collect perceptions and opinions as directly expressed by minors in their own words. The research teams and institutions working in this area include, for example, the OIA (Observatorio de la Infancia en Andalucía – the Childhood Observatory in Andalusia), an organ attached to the Dirección General de Infancia y Familias (Childhood and Family Affairs Office) of the Andalusian Regional Government and run by the FASS (Fundación Andaluza de Servicios Sociales – the Andalusian Social Services Foundation). The Observatory, which carries out research, training and documentation initiatives, has also implemented an information system whereby issues related to rights and assistance for under-18s can be properly identified, analysed from a technical point of view, monitored, addressed and made known to the public. In 2008, the OIA carried out a qualitative study which analysed what minors thought about their websurfing experiences (OIA, 2008). The results showed that minors consider the internet mainly as a tool for having fun, meeting people and accessing and consuming information, and that they use it above all for communicating with their peers via messaging. This mirrored the Spanish results from the EU-Kids Online project (Garitaonandía and Garmendia, 2007; Garmendia *et al.*, 2009), which also indicated that for minors the internet is a source of entertainment, relationships and information. Children were not, however, found to be prosocial online, since most of them do not engage in activities associated with solidarity, civic awareness or politics. A study combining both methodologies was also carried out by INJUVE (Instituto de la Juventud – the Youth Institute). Case studies were first made of 50 subjects, to enable the creation of a closed instrument capable of offering quantitative data on real situations experienced by adolescents. Another report, *Adolescents and Young People on the Internet: Opportunity Factors* (Rubio, 2009) was based on the answers given by 1111 individuals with ages ranging from 15 to 25. Here the most interesting results concerned the significance of the internet for adolescents and young people and the opportunities it offered them. In terms of significance, they considered the internet to be: (1) a means of communication (44.2 per cent); (2) a tool (32.9 per cent); (3) something indispensable (32.8 per cent); (4) a source of unlimited information (27.6 per cent); (5) a symbol of modernity and progress (24.1 per cent); (6) a source of opportunities (15.5 per cent); and (7) to a lesser degree, a danger (3.3 per cent) or something useless (1.9 per cent). The nine principal opportunities were: (1) communication with the world (66.7 per cent); (2) unlimited information (61.1 per cent); (3) a way of finding work (46.4 per cent); (4) entertainment (42.7 per cent); (5) unlimited relationships (36.7 per cent); (6) a channel for giving your opinion (18.5 per cent); (7) a feeling of involvement in something (10.9 per cent); (8) for cybersex (8.8 per cent); and (9) for privacy (8.5 per cent).

In 2010, the OIA (2010) published a report on activities and uses of ICTs by a sample group of Andalusian children. It identified minors as the driving force behind technological change. The main aim of the study was to discover what Andalusian minors thought about the way computers, the internet and mobile phones were habitually used in the 9–16 age group. A total of 1212 individuals from 45 public and state-subsidized private primary and secondary schools took part. Following the same methodology as INJUVE, the instrument employed was a questionnaire based on the results of earlier qualitative studies (OIA, 2008) and on other nationally and internationally recognized research instruments.

The activities engaged in by the children while online (OIA, 2010), listed in order of greatest frequency, were as follows: (1) using the internet to do school homework (94 per cent); (2) playing games online (85.7 per cent), visiting music web pages (83 per cent), reading or writing electronic mail (82.6 per cent); (3) downloading music, movies, videos (81.7 per cent); (4) chatting via instant messenger programmes (81.9 per cent); (5) sharing photos with friends (79.4 per cent); (6) reading news (78.4 per cent); (7) using the internet for hobbies (75 per cent); (8) downloading games, programs (72.4 per cent); (9) searching for information about films (70.1 per cent); (10) taking part in forums (47.2 per cent); (11) searching for information about health (42.1 per cent); and (12) chatting in public chat rooms (31.3 per cent). Apart from these activities, those questioned also mentioned the following uses for the internet: it helped them do school classwork (94 per cent); it informed them about what was happening in the world (82.5 per cent); and it helped them to pass the time (79.1 per cent). The internet also offered young people the chance to share activities with the family. For example, 50 per cent of those questioned said they used the computer with their family, and 37 per cent said they browsed the internet in the company of members of their family.

Taken globally, these results differentiate three types of macro-activity which can be carried out on the computer: (1) playing, listening to music and looking at photos; (2) studying or doing work for school; and (3) connecting to the internet. The activity which they devoted the most time to was connecting to the internet, followed by playing, listening to music, etc., with school work in last place.

With regard to mobile phone use, the activities engaged in most by the minors were: (1) communicating with members of their family (89.8 per cent); (2) taking photographs (88.8 per cent); (3) talking to friends (82.2 per cent); (4) recording their friends on video (65.9 per cent); (5) using the mobile phone to say hello to friends (50 per cent); and, finally, (6) downloading music and/or videos (11.7 per cent); and (7) connecting to the internet (6.7 per cent).

Again, no prosocial activities are found among the many uses made of the internet, although the minors did mention that they frequently participate in this type of activity in the 'real' world, for example, through weekend involvement in associations, NGOs, etc. (10.8 per cent) or helping out at home (92.5 per cent). It was therefore concluded that the internet is not usually employed for prosocial activities, but most boys and girls consider it a very positive tool and feel comfortable with the way they surf the Web.

Prosociality on the internet: a qualitative study

In the absence of research projects addressing the prosocial use of ICTs, we carried out two small pilot studies which provided information about whether young people use the internet in a positive manner and, more specifically, whether they use it for prosocial activities.

In the first study, a sample of 34 Instructional Psychology students in the fourth year of the Psycho Pedagogic Studies, of which 26.8 per cent were male, were asked two open questions about the ways in which the internet can be used prosocially and in which of those ways they use it. The results showed that many of the students do not know what type of prosocial activities can be conducted using the new technologies, and only 38.2 per cent of them declared that they engaged in any kind of prosocial activity. We therefore decided to carry out a second study in which we first asked about positive uses of the internet before going on to specify prosocial uses. We also included a question about cyberbullying. The target group comprised 62 young people in their first year of teacher training. They were asked four open-ended questions about the positive ways they use the internet and mobile phones, which prosocial activities they engage in, and what they would do if they found themselves involved in a case of cyberbullying. Their replies were used to create classification systems, and the results were as follows. The young people declared that the positive uses to which they put the internet mainly involved information and communication. In all, 88.7 per cent of them said they used the internet to search for different kinds of information, while 80 per cent said they used it for social interaction, above all using social networks which allowed them to be in contact with their families and friends but also to make new friends. Other activities they said they engaged in were organizational tasks such as filling in applications or searching for information (27.4 per cent), seeking employment (17.7 per cent), reading the press (17.7 per cent), academic work (12.9 per cent) and leisure-related activities (11.3 per cent) such as listening to music, watching movies, etc.

The young people questioned also mentioned positive uses of mobile phones. Some 88.7 per cent cited communication with loved ones (family and friends) via calls or text messages, 37.1 per cent cited emergency calls or asking for assistance, 32.3 per cent use the different phone applications such as the alarm clock, games, the calculator, taking photos, etc.; 17.7 per cent mentioned the use of the mobile phone to be reachable, for example, when their parents are out, 14.5 per cent for help or to support someone in need, and 6.5 per cent for obtaining information. When asked specifically whether they used ICTs prosocially or not, 80.6 per cent answered yes and mentioned different activities. The most frequently cited was that of helping or supporting loved ones (53.2 per cent), followed by seeking help in cases of emergency (21 per cent), collaborating with an NGO or helping people in need (14.5 per cent), sharing things such as notes or resources (6.5 per cent) and creating blogs (6.5 per cent).

Finally, when asked what they would do if they found themselves involved in a case of cyberbullying, 61.3 per cent said they would report it to the relevant authorities, 40.3 per cent would tell their family or friends, 29 per cent would change their email account or the number of their mobile phone, 14.5 per cent would go offline or switch their mobile phone off, 9.7 per cent would restrict incoming calls or access by friends to their profile on the social network, 9.7 per cent would ask for help and 4.8 per cent would adopt other measures.

Spanish initiatives to combat cyberbullying

As mentioned in this chapter, most initiatives aimed at encouraging the correct use of the new information and communication technologies have focused on either encouraging their use as learning tools or on preventing their inappropriate use. This is exemplified in the Spanish programmes and projects implemented to prevent cyberbullying over the past four to five years. When analysing the efforts made to prevent and combat cyberbullying, three major areas can be identified. The first is the actual objective of the initiatives being implemented, the second is the entities sponsoring those initiatives, which are bound to have some effect on their scope and specific impact, and the third is the proposed measures.

The objectives of the programmes and initiatives

Most of the programmes that have been implemented are designed to prevent or react to situations of internet abuse. Online harassment is therefore the focal point and the main aim of their measures. One example of this type of initiative is the e-Asturias programme being implemented as part of the Internet and Family Project (Gobierno del Principado de Asturias, 2006). Here, parents and teachers are offered guidelines and advice to facilitate safer internet use by means of security filters, game evaluation guides, etc. Another type of initiative with this same objective is being implemented by the Regional Government of the Canary Islands (www.gobiernodecanarias.org), which has created a website to combat risky behaviour on the internet. It includes a section dedicated specifically to cyberbullying. Similarly, the Pantallas Amigas (Friendly Screens) initiative (www.pantallasamigas.net), which is considered one of the most comprehensive, well-established proposals to have been presented in Spain, promotes the safe, responsible use of the new technologies through the prevention of risky behaviour such as sexting, cybercrimes, techno-addiction, grooming and cyberbullying. A whole page on its website is dedicated to this type of conduct (Pantallas Amigas, 2004). But mention should also be made of another very interesting group of programmes which are not exclusively aimed at eradicating and preventing risky behaviour but which also attach maximum importance to encouraging the safe use of the new technologies. These include the Pantallas Sanas (Healthy Screens) (Gobierno de Aragón, 2010) initiative, implemented by the Regional Government of Aragon, which promotes health and sociability in ICT usage, and the 'Escuela Espacio de Paz' ('School: a Space for Peace') Plans run by the Andalusian Government, which promote peaceful co-existence and the peace culture in the school environment (Binaburo, 2009).

The sponsors

The second area of analysis concerns the entities which sponsor and support the programmes described above. Other problems which affect co-existence both inside and outside schools are tackled through initiatives mainly created by the schools themselves or by other educational organizations or organisms. However, programmes designed to combat cyberbullying come from the public and the private sector and are largely conceived in institutions of a very different nature. At the government level, for example, the Tic-Tac project is being implemented as part of the Spanish Ministry of Industry's *avanza2* programme (Ministerio de Industria, Turismo y Comercio, 2009), or the proposals being drafted at the Observatorio Estatal sobre la Convivencia (State Observatory for Peaceful Co-Existence), under the auspices of the Ministry of Education, to train teachers to handle cases of cyberbullying. The different offices of Spain's Ombudsman, at both the national and regional levels, now offer legal and educational guidance with regard to cyberbullying. Action by non-governmental organizations range from initiatives set in motion by NGOs with exclusively educational goals, such as the 'Pantallas Amigas' programme (Pantallas Amigas, 2004) and the programme developed by the 'SOSbullying' association, to the recommendations drafted by the Comisión de Libertades e Informática (Commission for Liberties and Computer Technology), an organization which defends the Protection of Personal Information and combats cyberbullying through the Prometeo Project (Comisión de Libertades e Informática, 2008). Lastly, initiatives have also been forthcoming from the private business sector. Quite naturally, considering the high revenue represented by ICT consumption among the younger generation, mobile telephone operators and computer companies have implemented initiatives designed to contribute to the safe use of the internet and to prevent cases of online abuse and harassment. Microsoft has a whole website dedicated to the safe, responsible use of the internet, with a whole section providing information about cyberbullying and offering a series of guidelines regarding possible measures that can be taken to prevent it. The 'Customers' section on Vodaphone's web page has a code of conduct backed by all the mobile phone operators with specific information about cyberbullying and advice for parents and teachers on specific preventive measures (security packs and content filters). And Telefónica, perhaps the largest and most important of all Spain's telecommunications companies, also has a website dedicated to the responsible use of new technologies.

Proposed measures

Generally speaking, it can be concluded that most of the programmes analysed are essentially informative. But projects have also been undertaken, although much less frequently, which have developed specific teaching materials, have set up internet or telephone-based helplines, have contributed to the training of teaching staff and families or have encouraged the participation of young people. Each of these approaches is examined in more detail below.

Information

To fully understand a phenomenon like cyberbullying and address it appropriately with suitable tools, it is necessary to recognize its nature, identifying its features and complexity. The first step in detecting and preventing cyberbullying is therefore to know exactly what it is, in what form it appears, and to what extent it can have a psychological and social impact on those involved. In all the initiatives implemented in Spain, information is the first tool to be used in combating the problem. Moreover, the support given by these projects means that the information made available is usually quite extensive and detailed. Informative websites include that of Pantallas Amigas (www.pantallasamigas.net), that of the independent 'SOSbullying' association (www.sosbullyingespana.es), and that of the Protégeles association (www.protegeles.com).

Many of these websites also have forums where opinions can be shared, suggestions and procedures can be put forward and cases of cyberbullying can be reported. Such forums exist on the Protégeles and SOSbullying sites and on those of most offices of the Child Protection Ombudsman. The office of the Child Protection Ombudsman in the Basque Country, for example, offers guidance in the field of electronic security (www.ararteko.net/RecursosWeb).

Information is also provided in projects implemented by schools, which are in a position to ensure that pupils know that the information is available, and can facilitate access to it. Examples can be found in many schools throughout the country. For example, the IES Ramiro II secondary school in La Robla, in León, publishes interviews and news items related to cyberbullying, while the IES Almería has created a wiki in which the teachers have included useful information on preventing and detecting cyberbullying along with other online resources and useful links that can be visited by parents and pupils.

Teaching materials

The creation of teaching materials complements the provision of information. These materials usually include practical suggestions on how to use the new technologies safely and responsibly or on how to approach cases of cyberbullying with sons, daughters or pupils. Of particular interest are the guides produced by the e-Asturias programme in its Internet and Family Project and the material created by Pantallas Amigas, which uses multimedia content on the subject of cyberbullying and an educational video game that can be played on mobile phones to encourage the safe, responsible use of the internet. The Prometeo programme developed by the Comisión de Libertades e Informática (Commission for Liberties and Computer Technology), an organization dedicated to the protection of personal information, has also produced guides designed to encourage children and adolescents to use the internet safely (www.asociacioncli.es).

Awareness-raising videos are another essential teaching resource for combating this type of problem. Examples include the informative videos produced by the Protégeles association (www.protegeles.com) and the awareness campaign conducted by the Regional Government of Madrid's Child Protection Ombudsman (www.defensordelmenor.org), which, as well as launching this initiative, has also collaborated with different child protection NGOs (including the Protégeles association mentioned above).

Helplines

Helplines are a good tool for identifying specific cases of online abuse and harassment. The availability of well-informed anonymous support capable of tackling the problem in all its complexity (the legal, psychological and educational aspects) is instrumental in bringing cases out into the open. But finding a good helpline is not easy. At present, and (in our opinion) despite the very recent implementation of these programmes and projects, a number of helplines exist. Some, like the ones set up by the Regional Government of Asturias (e-asturias) and the Regional Government of Valencia's Plan PREVI (www.edu.gva.es/eva/es/previ), are run by public administrations, while others are managed by non-governmental organizations. The latter include the cyberbullying helpline run by the Protégeles association (www.internetsinacoso.com), which also offers teachers and parents advice, information, legal guidance and downloadable documentation with which to report cases to the relevant authorities.

The training teachers and families

As a line of action, the training of those involved in educating children stems largely from the initiatives implemented by the education authorities at national, regional and local level. This type of training has a relatively short history. In fact, the issue of cyberbullying was only added to the agenda of the teacher training sessions organized by the State Observatory for Peaceful Co-existence in 2009. Similarly, only recently have regional and local government organisms started to provide specific training for teachers and/or families in ways of addressing the problem of cyberbullying. Attention should be drawn here to the initiatives conducted by the Andalusian and Valencian education authorities, which have focused on training teachers to handle cyberbullying, and the cyberbullying training programmes set up by the AAPS (Asociación Aragonesa de Psicopedagogía – the Aragonese Association of Psychopedagogy: http://www.psicoaragon.es). Non-governmental associations also provide training for teachers and families. The Pantallas Amigas association mentioned above organizes informative symposia and workshops, while many schools and colleges run their own training programmes (for example, several of Spain's secondary schools: see the webpage of the IES Gran Capitán school in Córdoba at www.iesgrancapitan.org).

Participation by young people

Involvement of the young in the projects and initiatives under discussion is stimulated mainly by regional and local education entities, particularly schools. Examples can be found in the different awareness-raising symposia and gatherings in which young people appear as presenters, describing their own experiences of internet victimization. Testimonies of this type have a great impact on their audience because they bring not only the young people themselves but also their parents much closer to a social reality of a virtual 'cyberworld', the impact of which is often underestimated or trivialized. The Youth Meetings which have been organized by the Valencian Regional Government since 2006 are a good example of how young people can share ideas and their experiences regarding their preoccupations at school (see http://jmada.eresmas.net/III%20Encuentro/iiii_encuentro%201.htm).

Likewise, at the local level, schools (above all secondary schools) have been implementing educational projects which involve pupils in the creation of virtual newpapers. This encourages the positive use of the new technologies while at the same time serving as a channel through which any cases of school violence can be identified and reported. Pupils are also involved in blogs, discussion forums and even in the making of videos.

Conclusion

It cannot easily be confirmed that Spain's schoolchildren and young people are fully aware of how to use their internet skills and abilities prosocially. In fact, perhaps in general it is difficult to be aware of the nature of their prosocial behaviour in everyday life. Prosocial behaviour has become associated with the unusual, with acts of heroism, while behaviour online has become just another part of everyday life. However, everyday life is full of acts which cross the norms of accepted behaviour. There exists a broad range of positive actions which, when carried out, are beneficial both to their author and to others. According to the theory of positive psychology (Seligman, 1991), it is optimism – a positive, understanding willingness to be happy – that separates a situation of risk from a situation in which personal development and learning are possible. But this is different from the extra effort that is needed to go one step further than the demands of normal behaviour, which seems to have been identified as prosocial conduct. In spite of the fact that the internet is being used in numerous situations to give support, company and help of all kinds, our young people do not seem inclined to use it for purposes which go beyond the limits of comfortable, conventional behaviour. Admittedly, we also believe that this attitude extends to their offline conduct. As we have seen in our small studies, our schoolchildren and young people consider the positive use of the new technologies mainly in terms of the opportunities it offers for seeking information and as an aid to academic work.

Society has tended to react to the way the young employ the new technologies with alarm rather than welcoming the positive uses of such a medium. Initiatives launched to encourage ICT's use have focused very much on alerting users to the negative use of the same, as is exemplified by programmes dedicated to the prevention of cyberbullying. In Spain, these initiatives have mainly been initiated by private entities with positive attitudes. In general, they may therefore be considered prosocial initiatives because their aim is to avoid difficulties and prevent problems or abuse between users. Initiatives of this type, some of which are government-sponsored, are educational in nature and are geared towards ensuring a more efficient, positive use of technology: e-safety. It could be said that what is needed in Spain is greater emphasis on the positive, prosocial use of the ICTs. At the very least what is required is the raising of users' awareness regarding the positive use which may still now be possible in an online pattern of behaviour. This is particularly important because online behaviour is beginning to mirror offline behaviour in terms of its social functions and degree of importance for social interaction. We think that understanding the psychological and social dimensions of cyberbehaviour is an important research challenge which has still not been addressed in sufficient depth.

References

Binaburo, J.A. (2009) *Escuela Espacio de Paz*, Seville: Consejería de Educación de la Junta de Andalucía.

Comisión de Libertades e Informática (2008) *Proyecto CLI–PROMETEO: Manual Práctico de Uso de las Tecnologías de la Información*, Madrid: Web y Media Diseñadores S.L.

European Commission (2008) *Eurobarometer Survey: 'Towards a Safer Use of the Internet for Children in the EU: A Parents' Perspective'*, available at: http://ec.europa.eu/information_society/activities/sip/eurobarometer/ (accessed 3 November 2010).

Garitaonandía, C. and Garmendia, M. (2007) *Cómo Usan Internet los Jóvenes: Hábitos, Riesgos y Control Parent*, available at: http://ehu.es/eukidsonline/INFORME%20FINAL-INTERNET.pdf (accessed 11 June 2008).

Garmendia, M., Garitaonandia, C., Casado, M.A. and Martínez, G. (2009) 'Resultados empíricos Europeos', paper presented at first International Congress, 'Minors in ICT', Gijón, October.

Gobierno de Aragón (2010) *Pantallas Sanas*, Zaragosa: Gobierno de Aragón.

Gobierno del Principado de Aturias (2006) *Internet and Family Project*, available at: http://internetyfamilia.es/ (accessed 11 June 2008).

INJUVE (2004) *Jóvenes, Relaciones Familiares y Tecnologías de la Información y la Comunicación*, Madrid: Consejería Técnica de Planificación y Evaluación.

Livingstone, S. (2009) *Children and the Internet: Great Expectations and Challenging Realities*, Cambridge: Polity.

Livingstone, S. and Haddon, L. (eds) (2009) *Kids Online: Opportunities and Risks for Children*, Bristol: The Policy Press.

Ministerio de Industria, Turismo y Comercio (2009) *Plan avanza 2*, Madrid: Gobierno de España.

OIA (2008) *Nuevas Tecnologías e Infancia y Adolescencia*, available at: http://www.juntadeandalucia.es/observatoriodelainfancia/oia/esp/descargar.aspx?id=1969&tipo=documento (accessed 1 September 2010).

——(2010) *Informe sobre Actividades y Usos de TIC entre los Chicos y Chicas en Andalucía*, available at: http://www.juntadeandalucia.es/observatoriodelainfancia/oia/esp/descargar.aspx?id=2734&tipo=documento (accessed 1 September 2010).

Pantallas Amigas (2004) *Por un Uso Saludable*, Bilbao: Edex.

Rubio, A. (2009) *Adolescentes y Jóvenes en la Red: Factores de Oportunidad*, Madrid: Instituto Nacional de la Juventud.

Seligman, M. (1991) *Learned Optimism*, New York: A. Knopf.

Simone, R. (2000) *La Tercera Fase:Formas de Saber qué Estamos Perdiendo*, Madrid: Taurus.

Chapter 7

Using the internet positively in schools

The case for notebooks

Vera Popper, Dagmar Strohmeier and Christiane Spiel

By now there is much evidence that the internet can be a threat for children and youth (e.g. Gradinger *et al.*, 2009; Smith *et al.*, 2008) with recognition that efforts should be undertaken to protect them. In contrast, less research has been undertaken to examine the positive use of information and communication technology (ICT). Recent findings, however, have demonstrated the internet's potential for learning and enhancing social relations (Guan and Subrahmanyam, 2009). In particular, there is evidence that adolescents use the internet mainly to strengthen relationships they have in real life (Subrahmanyam and Greenfield, 2008).

When used as an educational tool, ICT has the potential to change teaching and learning in schools. In this chapter, we present an evaluation of the project 'E-learning and e-teaching in notebook classes' ('The Notebook Project'), which established the electronic notebook as a positive example of ICT use in school contexts. This Austrian project started in 2000 with support from the Federal Ministry for Education, Science and Culture, but with neither clear goals nor defined criteria for success. In 2002, almost 4000 students from 170 classes had participated in the project nationwide. As a consequence, the Austrian Federal Ministry commissioned an evaluation to determine whether the project had had any positive effects. This chapter describes the Notebook Project as it was established in 2002 before its evaluation, with the main focus on the evaluation model. It describes an exemplar which illustrates how to retrospectively define the goals of such a project, and how subsequently to evaluate its effectiveness and implementation. The notion of programme evaluation is briefly described. Finally, recommendations for the use of ICT as an educational tool are given.

The project 'E-learning and E-teaching in Notebook Classes'

In supporting the use of ICT in school contexts, the Austrian Federal Ministry for Education, Science and Culture intended to foster self-regulated learning (Dorninger and Horschinegg, 2002). A new culture of learning, one comprising, more 'construction' of knowledge instead of 'instruction', was to be established. This new culture of learning should consist of three elements, which complement each other: (1) learning via presentations from teachers; (2) self-regulated learning via sources such as books or multimedia-based contents; and (3) interactive, cooperative learning via ICT in company with other learners. Notebook classes were seen as one possibility to change the culture of learning. Because the Notebook Project was a school-driven initiative, and therefore developed somewhat 'auto-didactically', neither standardized goals nor clear criteria for success were formulated. These notebook classes had one element in common: that notebooks should be used as the main resource for teaching and learning in all subjects. Consequently, all students in the schools which participated were equipped with notebooks at the parents' cost. At the time that the evaluation study was commissioned, pioneer schools had had notebook classes for two years with other schools having just started their classes: in sum, 170 classes in four different school types. Our study sought to investigate the following two questions: (1) What skills and competencies are promoted by learning via notebooks? and (2) What are the special features of successful notebook classes?

Programme evaluation: a brief introduction

In general, evaluations are concerned with the effectiveness of programmes, projects or interventions (Rossi *et al.*, 2004; Spiel, 2005). The process of evaluation is characterized by certain activities or tasks: defining goals and indicators for measuring them; setting standards of effectiveness; designing the evaluation; collecting and analysing information and reporting results. One important characteristic of evaluations that is useful for the clients (the individual, a group or an organization that is commissioning the evaluation) is that evaluation goals should be defined by the clients (Patton, 1996) with the support of the evaluator. This is the first step to engage the important stakeholder groups in the process of evaluation, and is important as participation increases acceptance of the results and a willingness to change programmes according to the results of an evaluation (Guba and Lincoln, 1989). Stakeholder identification is also described as one important Evaluation Standard (Gesellschaft für Evaluation, 2008) in the group of utility standards, which aim to ensure that an evaluation is guided by both the clarified purposes of the evaluation and the information needs of its intended users.

At the very beginning of every evaluation, the first thing to look at are the two levels of objectives: (1) the objectives of the evaluation; and (2) the objectives of the programme being evaluated (Spiel *et al.*, 2010). Optimally, the evaluators are integrated into the process of defining the objectives of the measure from the beginning. Therefore, a joint explication of objectives, preferably in the context of a workshop with the most important stakeholders, has to be conducted, in which the following questions need to be clarified:

1 What objectives does the programme intend to achieve?
2 How will these objectives be achieved?
3 How well does the programme correspond with the objectives?
4 How can the achievement of the objectives be recognized?
5 What will be different after the programme has been carried out?

On this basis, questions are formulated which the evaluation is expected to answer.

Evaluation of the 'Notebook Project'

According to the general concerns of programme evaluation and the very vague goals of the Notebook Project, we designed an evaluation model which consisted of the following three parts: (1) identification of common project goals and evaluation criteria; (2) evaluation of the effectiveness of the Notebook Project according to the defined goals; and (3) analysis of the implementation of the Notebook Project. Results are used as basis for recommendations which are described in the final section of this chapter. All three evaluation parts are described in the same way: our intentions, the method we used, and exemplified results.

Identification of common goals and evaluation criteria of the Notebook Project

Intention

As previously mentioned, neither standardized goals nor clear criteria for success were formulated for the Notebook Project when it was instigated. Therefore, our first task was to clarify our goals and the evaluation criteria. Furthermore, we were interested in gathering basic information about the organization and didactics/pedagogies of the project in general.

Method

A two-day workshop with 13 systematically selected school teachers and principals, as experts on the Notebook Project, was organized and facilitated. As specifically as possible, the ways in which the success of the Notebook Project could be identified were itemized, and indicators were defined which sought to distinguish notebook students from students taught by other commonly employed methods. Selection criteria for workshop participation were, for example, school type and duration of participation in the pilot project. Some teachers were selected to take part in the workshop because they were not supportive of the project. It was very important for the credibility of the evaluation to take into account the arguments of those who were convinced of the negative side effects of the notebook usage.

To retrospectively identify which skills and competencies, so-called 'interdisciplinary key qualifications', are promoted by e-learning via notebooks, we used workshop techniques such as brainstorming or small group discussions. To collect relevant criteria for a successful implementation of notebooks classes and relevant criteria for didactics/pedagogies, we used mind mapping.

Results

The workshop participants agreed that notebook learning should particularly promote specific skills and competences. Consequently, as proposed by the participants in the workshops: notebook students should outperform traditional students in the following competences: (1) information management: 'they are better at solving problems in the area of information management'; (2) team competencies: 'notebook students work more often in teams and are able to do so more professionally than students from traditional classes'; and (3) self-organization, self-efficacy and motivation for learning in school.

Workshop participants also identified that the following potential risks would be more likely for the group of notebook students: (1) decrease in concentration competence; (2) addiction to the notebook, for example, computer games, chats, internet use; and (3) drop in school grades.

The mind mapping exercise identified several relevant criteria for the organization, preparation and implementation of, and teaching in, notebook classes. Criteria deemed relevant for organization were, for example, appropriate technical equipment, teachers' computer skills levels, and the availability of relevant digital content. To measure didactics and pedagogies, indicators such as giving continuous feedback, defining subordinate targets and transparent assessment criteria were defined.

These clarified goals and evaluation criteria subsequently formed the basis for the data collection instruments in parts two and three of the evaluation study.

Evaluation of the effectiveness of the notebook classes

Intention

The second part of the evaluation investigated whether the goals defined in the workshop had been achieved by the Notebook Project, i.e., the defined skills and competences promoted via notebook learning were to be measured concretely and the suspected risks were to be confirmed. To get a fit between the very practical Notebook Project and the measurement instruments we decided not to use academic measures but to develop and use ecologically valid instruments. Furthermore, as the project was already ongoing when the evaluation was commissioned, there was no baseline data available for comparison. Consequently, we had to use a treatment–comparison evaluation design.

Method

1 *Samples*: Data collection was done in 25 systematically selected notebook classes and control classes. Selection criteria were school type, federal state and duration of participation in a notebook class (one year/two years). Some 490 students participated (76.1 per cent notebook students, mean age = 16.7 years, 41 per cent female) in the performance test that sought to measure the interdisciplinary key qualifications (skills and competencies) as defined by the workshop participants.

2 *Instruments*: In the following, we describe only some of the instruments used in detail. More information about the whole test inventory can be found in the project report (Spiel and Popper, 2003) as well as in Popper and Spiel (2010). A new test inventory with high ecological validity was developed to investigate information management (search, selection, process, and reproduction of information) and team competencies in students. To collect data on self-organization, self-efficacy and motivation, well-established questionnaires were used.

2.1 *Information management*: The students were asked to compile a work sample based on the following written instruction:

> A school class from the United States, consisting of 20 students and their two teachers, is coming to your school for a two-day visit as part of their trip through Austria. You are asked to organize this visit, so that the guests can enjoy their stay. With the help of three of your colleagues you are to organize these three agenda topics: (1) the school class must be picked up at the train station on Friday, 12.45 p.m.; (2) planned activities are the visit to a theatre or musical; and (3) sightseeing.

The students were asked to organize the event, first, without any concrete instructions of how to solve the task (recognition of information). The aim of the first part of the task was to measure self-regulatory outlining solution strategies. Students' solution strategies were rated as follows:

- *Task assignment*: Were the tasks divided between the three helping colleagues? Who does what?
- *Time planning*: Which tasks need to be done earlier? Is there an assignment of priorities?
- *Activities*: Were all three activities mentioned?

Second, the organizational task was presented again, this time with concrete instructions of how to solve it (process and ability to reproduce information).

Students were asked: (1) to prepare a project plan and to share the work between all the involved colleagues; (2) to document their internet search; (3) to create a PowerPoint presentation to introduce the planned activities to the head of their school; and (4) to write an e-mail in English to the American students to inform them about the planned activities.

Students' products were rated as follows: (1) Project plan: have task assignment and time planning been taken in consideration? Is the plan realizable? (2) Internet research: is there a link list? (3) Presentation: is it appropriate for the target audience, the head of the school? Is the amount of information enough/too much/too little? Is the layout well done? and (4) e-mail: is it written in English (correct/basic/elaborate English)? Are all the planned activities mentioned?

2.2 *Team competencies*: Due to the constraints of financial resources, it was not possible to investigate the students' real behaviour while team working. Therefore, we sought the relevant criteria of successful team work. Students were asked to write down: (1) the relevant attributes of a team member for successful team work; and (2) the relevant criteria of the whole group's working progress in order to realize successful team work. The number of reasonable attributes as well as relevant criteria in terms of the literature, such as a sense of responsibility or commitment (e.g. Schuler, 2004), were counted in both tasks.

2.3 *Self-organization, self-efficacy and motivation*: To collect data on these constructs, well-established questionnaires were used (Rost and Lamsfuß, 1992; Jerusalem and Satow, 1999; Koeller *et al.*, 2000; Schober, 2002; Spiel and Schober, 2003).

2.4 *Potential risks of using the notebook*: To examine whether students showed signs of addiction to the notebook, we asked the students to estimate their hours of computer use at school and at home. Additionally, they were asked to describe their most common computer activities for both domains (e.g., internet research, homework, chats, or computer games). Concentration competence was tested with the d2-test (Brickenkamp, 2002), which measures velocity and accuracy of crossing out specially marked letters. Finally, we collected data about students' grades from the final year of a lower grade until the current school report, to investigate the hypothesized decrease in school grades.

The whole test inventory was pre-tested for comprehensibility and required time to complete the tasks.

Data collection

All students used a computer to complete the tasks. Notebook students used their own notebook, students in the control classes used a computer provided by the school in a laboratory. To avoid system bias, namely not being used to working with computers, we asked the students in the control classes if they owned a computer at home. Surprisingly, only two students in the control class did not own one. Therefore, each participant should have been accustomed to working with a PC. Questionnaires for measuring self-organization, self-efficacy, and motivation as well as the d2-test were given in a paper-and-pencil version. The data collection took place in the students' school, and was conducted by trained members of the Vienna Pedagogical University.

Results

First, a brief overview of the goals which could be achieved is reported. Second, three results of our evaluation study are presented in detail, because they are the most interesting ones in the context of this book: teamwork, self-efficacy, and usage of the computer.

Students in notebook classes (for more details, see Popper and Spiel, 2010):

- were more capable of working out a solution independently (*information management* – recognition of information);
- could solve an organization task better (*information management* – process and reproduce information);
- specified more relevant criteria for *team work*;
- showed higher scores in *self-efficacy*;
- did not differ from students of 'traditional' classes used as control groups with respect to *learning motivation*, *self-organization* and how much they like school.

Considering *potential risks*, notebook students showed *no* higher use of the computer at home, the same performance in the concentration test, and no difference in their school grades (same mean grades and same decrease in grades over the years) in comparison to students from control classes.

One major concern related to computer use in general or internet use in particular is that youth would neglect their relationships with peers in real life and face-to-face communication in class would decrease. Teachers from notebook classes participating in our workshop, however, argued that working with a portable computer facilitated working in groups. Also, because additional information is more easily available, teachers could give their students broader tasks with subordinate targets to be solved in teams. Therefore, notebook students should be more experienced in working in teams and have higher team competencies. Our data supported this hypothesis: that notebook students were able to specify more relevant criteria for teamwork ($F(1,481) = 15.2, p < 0.001, \eta_p^2 = 0.03$); $M = 1.8, SD = 0.7$ vs. $M = 1.5, SD = 0.7$) that corresponded to the literature. Of course this effect is not caused by the computer itself, but by different didactics/pedagogies which are supported by the notebook. Interestingly, all students mainly named criteria concerning the social aspects of teamwork. The most infrequent criterion was 'goal-setting'.

Notebook students were also found to show higher scores in self-efficacy ($F(1,478) = 7.1, p < 0.01, \eta_p^2 = 0.02$); $M = 4.3, SD = 0.6$ vs. $M = 4.1, SD = 0.8$; scale 1–6 with 6 = highest agreement) than students from control classes. One possible explanation for these differences, though small, could be that teachers in notebook classes facilitated more positive experiences that fostered students' self-efficacy, e.g., letting students find strategies to solve problems themselves, than in control classes. It is also possible that notebook students discovered their superior skills in handling computers, as compared to their teachers, highlighting a change of roles between student and teacher. Differences between girls and boys in self-efficacy and attribution were also reproduced, supporting prior research (e.g. Spiel and Schober, 2003). However, our data did not support findings that computer and internet use fostered motivation for learning (Rau *et al.*, 2008) – at least not learning at school.

Concerning the risk of addiction to the notebook, no evidence was found in the data. The mean time of using the computer at home did not differ significantly between the two groups: on average, all students spent 2.5 hours per day, mainly doing homework, playing games and surfing the internet (Figure 7.1). The mean usage time in school was much higher for the notebook students (M = 5.5 hours, SD = 1.9) than for the others (M = 0.9, SD = 0.8). It must be mentioned that the time spent in front of a monitor, all in all, per day is therefore quite long for students from notebook classes with a high variation between individuals (M = 8.3 hours, SD = 7.5).

As expected, results showed differences between notebook classes in realizing the defined goals: 50 per cent of the 20 students with the best test scores concerning information management and team competencies came from the four best notebook classes, whereas 3.3 per cent would be expected randomly (χ = 22.76, p < .01). Considering these findings, the aim of the third part of the evaluation was to identify features relevant for the success of the Notebook Project.

Analysis of the implementation of notebook classes

Intention

In the third part of the evaluation we intended to retrospectively identify how the implementation of the notebook classes was established. In effect, we wanted to know whether implementation and teaching in notebook classes that are successful according to the defined goals (Part II performance test), systematically differ from less successful notebook classes.

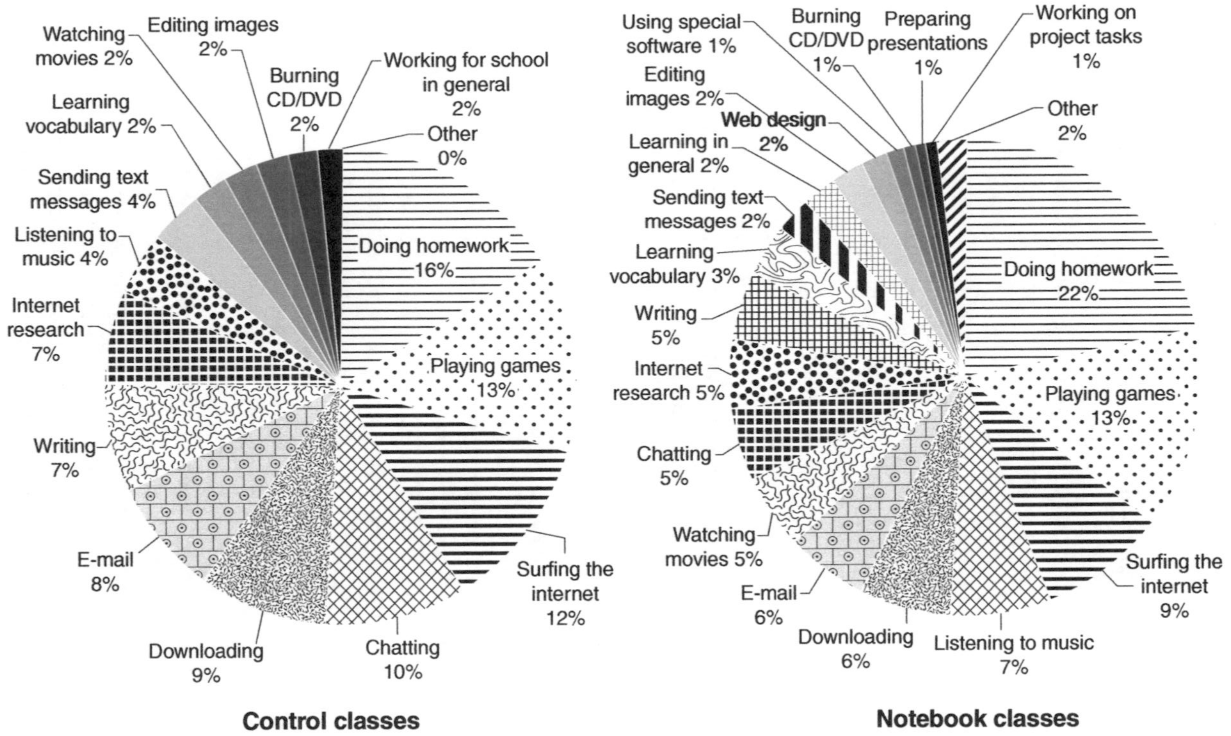

Figure 7.1 Computer use at home

Method

An extreme group comparison as an evaluation model was applied. Teachers and students from the best and poorest classes concerning information management and team competencies were investigated in detail.

1 *Sample*: In sum, 137 notebook students from eight classes (six of the best notebook classes and two of the poorest, all school types represented) and 37 of their teachers (persons in charge of the notebook project per school, form masters and teachers of major subjects) were included in the second data collection.
2 *Instruments*: On the basis of the workshop results concerning organization of the Notebook Project and the didactics/pedagogies involved, a structured telephone interview was developed to identify differences between the activities of the best and poorest classes. Interviews with teachers lasted from 20–60 minutes and included mainly open-ended questions concerning implementation of the notebook class, concrete descriptions of lesson preparation, teaching in class and grading. Students were asked to fill out a questionnaire, describing the activities they carry out mainly via their notebook in school and at home, in detail, their perception of changes in teaching and their own learning, and their reason for choosing a notebook class.

Results

Results showed high similarities between teachers' and students' attitudes towards teaching and learning in notebook classes in both groups, but systematic differences between the groups. In successful notebook classes, both teachers and students took over perspectives of the other's role and therefore tried to support each other, but this was not the case in less successful classes. Teachers from those classes primarily showed external attribution of problems (e.g., technology, lack of intelligence). Teachers of the best classes rated their students' motivation for learning higher than it actually was and higher than their colleagues from poor classes did.

There were no significant differences between the described students' activities at home and at school, but students from the poor classes in relation to the performance test, reported playing computer games and downloading music or movies from the internet more often than did students from the best classes. Additionally, students from poorer classes did not notice changes in teaching or in their own learning. Teachers from those classes reportedly used the notebook as an 'advanced typewriter' more often than their colleagues from best classes did. There were no systematic differences in the students' reasons for attending a notebook class.

Concerning the organization of notebook classes, the main result was the importance of an appropriate preparation for implementation (e.g., information, enough time, clarification of technical issues). In sum, no single factor of success could be identified, but numerous relevant features concerning organization and teaching were found.

Based on these evaluation results and our own research, a catalogue of recommendations was formulated, which should support both the implementation of notebook classes and the continuous improvement of teaching and learning in notebook classes.

Recommendations for positive uses of the internet: lessons learned from the evaluation of the Notebook Project

In concluding, recommendations for organizational topics as well as for teaching are presented.

Preparation and implementation

Before starting a notebook class, an internal discussion is needed: the head of the school and teachers together should define: (1) the goals the school wants to achieve with the notebook class; (2) the design didactics; and (3) nomination of the teachers. Also, further training for teachers has to be organized and technical details and support arranged. Our data showed that giving this communication process enough time (up to one year) and engaging all relevant stakeholders at an early stage are critical success factors in this phase.

When informing and preparing the students, it is crucial to define rules for using the PC, the internet, or other portable electronic devices (e.g., mobile phones with internet access) at school. For example, some of the notebook classes experienced positive effects by having the students sign contracts, where the rules and also the objectives of the notebook class were listed. It is also important to outline the disciplinary consequences, especially if another person is harmed (e.g., cyberbullying). To enforce information about the consequences of cyberbullying for the victim, researchers recommend showing students hard-hitting video clips (Hinduja and Patchin, 2009).

To meet the technical challenges, some schools took advantage of so-called 'Junior System Administrators', talented and interested students, who support teachers with technical tasks especially in the implementation phase. These students both received training and trained their teachers, which led to the first role shift and fostered learning for students and teachers.

Preparation of lessons, teaching and grading

As these results have indicated, information management can be successfully enhanced and more easily undertaken via notebooks. This competence is essential for many professions and in particular for any further university studies. In spite of the fact that young people are growing up with the internet and computers, internet research and judging the quality of retrieved information, correct citing, or working in groups (especially virtual groups) must be explained, explicitly taught and practised. According to prior research (Schober *et al.*, 2008), students only continued to use new tools and methods if they had already tried them, and if they were successful. It also became obvious from this study that it is crucial to support slower students when beginning to work with new methods, as they tend to struggle, not only with the tasks but also with technique. Some teachers in our study reported that the atmosphere and social relationships in class had been improved by team working, when good students are urged to support poorer students. This method not only enhanced social relationships, but also prevented the gap between good and poor students from widening. Teachers in our study recommended establishing team working even in the beginning phase of a notebook class.

In addition to easier organization of team work, the internet offers many opportunities to individualize teaching and learning, e.g., providing more or less advanced audio files for teaching languages according to different students' levels.

It is recommended that students are asked to shut their notebooks while teachers are explaining a task or the class is discussing a topic. This is particularly important in the first months with the new tool, because some teachers and students reported a decrease in offline communication in class at the beginning of the Notebook Project. Continuous feedback helps to ensure students' motivation, in spite of the additional work at the beginning. Teachers also reported a shift in their role from 'instructor' to 'coach'. This change in the relationship with the students resulted in an increase in their teaching motivation and also in a positive change of the atmosphere in class.

To assess the students' work, teachers in our study used broader tasks in order to avoid copy and paste strategies. Instead of questions with only one correct answer, their students must explain how they found a solution or must write down the approach to the problem. In general, to benefit from the positive use of ICT in school education, the tool should always follow and support the intended goals of a lesson and not be an end in itself.

Conclusion

The goals of the present chapter were twofold: (1) to show a positive example of ICT use in schools by presenting the successful Notebook Project; and (2) to show the development of an evaluation study under challenging conditions, which meets the defined standards (Gesellschaft für Evaluation, 2008) with a specially tailored and ecologically valid test inventory, when explicit goals of the programme had to be defined retrospectively. If the described recommendations could be included in the use of ICT as an educational tool, successful learning should be achieved.

References

Brickenkamp, R. (2002) *Test d2 – Aufmerksamkeits-Belastungs-Test* [Test d2–Attention-Resilience Test], 9th edn, Goettingen: Verlag für Psychologie Hogrefe.

Dorninger, C. and Horschinegg, J. (2002) 'Mobile e-Learning und e-Teaching. Ein Modellprojekt mit Schülernotebooks an Österreichs Sekundarschulen' [Mobile e-learning and e-teaching: a model project with student notebooks in Austrian secondary schools], *Die deutsche Schule* [The German School], 94(2): 247–56.

Gesellschaft für Evaluation e.V. (2008) *Evaluation Standards*, available at: http://www.alt.degeval.de/calimero/tools/proxy.php?id=19084> (accessed 15 September 2011).

Gradinger, P., Strohmeier, D. and Spiel, C. (2009) 'Traditional bullying and cyberbullying: identification of risk groups for adjustment problems', *Journal of Psychology*, 217(4): 205–13.

Guan, S.A. and Subrahmanyam, K. (2009) 'Youth Internet use: risks and opportunities', *Current Opinion in Psychiatry*, 22(4): 351–6.

Guba, E.G. and Lincoln, Y.S. (1989) *Fourth Generation Evaluation*, Newbury Park, CA: Sage.

Hinduja, S. and Patchin, J.W. (2009) *Bullying Beyond the Schoolyard: Preventing and Responding to Cyberbullying*, Thousand Oaks, CA: Corwin Press.

Jerusalem, M. and Satow, L. (1999) 'Schulbezogene Selbstwirksamkeitserwartung' [Expectations of self-efficiency related to school], in R. Schwarzer and M. Jerusalem (eds) *Skalen zur Erfassung von Lehrer-und Schülermerkmalen* [Scales for measuring student and teacher characteristics], Berlin: Freie Universität Berlin.

Koeller, O., Schnabel, K.U. and Baumert, J. (2000) 'Der Einfluss der Leistungsstärke von Schulen auf das fachspezifische Selbstkonzept der Begabung und das Interesse' [The effect of schools' capacity on subject-specific self-concept of ability and interests], *Zeitschrift für Entwicklungspsychologie und paedagogische Psychologie* [Journal of Developmental Psychology and Educational Psychology], 32: 70–80.

Patton, M.Q. (1996) *Utilization-Focused Evaluation: The New Century Text*, 3rd edn, Thousand Oaks, CA: Sage.

Popper, V. and Spiel, C. (2010) 'Entwicklung eines komplexen dreistufigen Evaluationsdesigns unter schwierigen Rahmenbedingungen: die Evaluation von Notebook-Klassen' [Development of a complex three-stage evaluation design under difficult conditions: evaluation of notebook classes], *Zeitschrift für Evaluation* [Journal of Evaluation], 9(1): 7–28.

Rau, P.P., Gao, Q. and Wu, L. (2008) 'Using mobile communication technology in high school education: motivation, pressure, and learning performance', *Computers & Education*, 50(1): 1–22.

Rossi, P., Lipsey, M.W. and Freeman, H.E. (2004) *Evaluation: A Systematic Approach*, 7th edn, Thousand Oaks, CA: Sage.

Rost, D. and Lamsfuß, S. (1992) 'Entwicklung und Erprobung einer ökonomischen Skala zur Erfassung des Selbstkonzepts schulischer Leistungen und Fähigkeiten' [Development and testing of an economic scale for measuring the self-concept of achievement and abilities in school], *Zeitschrift für paedagogische Psychologie* [Journal of Educational Psychology], 6: 239–50.

Schober, B. (2002) *Entwicklung und Evaluation des Münchner Motivationstrainings (MMT)* [Development and evaluation of the Münchner Motivationstraining (MMT)], Regensburg: Roderer.

Schober, B., Wagner, P., Reimann, R. and Spiel, C. (2008) 'Vienna E-Lecturing (VEL): learning how to learn self-regulated in an internet-based blended learning setting', *International Journal on E-Learning*, 7(4): 703–23.

Schuler, H. (2004) *Organisationspsychologie: Gruppe und Organisation* [Organizational psychology: group and organization], Goettingen: Hogrefe.

Smith, P.K., Mahdavi, J., Carvalho, M., Fisher, S., Russell, S. and Tippett, N. (2008) 'Cyberbullying: its nature and impact in secondary school pupils', *Journal of Child Psychology and Psychiatry*, 49: 376–85.

Spiel, C. (2005) 'Program evaluation', in C.B. Fisher and R.M. Lerner (eds) *Applied Developmental Science: An Encyclopedia of Research, Policies, and Programs*, Thousand Oaks, CA: Sage.

Spiel, C. and Popper, V. (2003) *Evaluierung des österreichweiten Modellversuchs 'E-Learning und E-Teaching mit Schüler/-innen-Notebooks'. Abschlussbericht* [Evaluation of the Austrian-wide model test 'E-learning and e-teaching with student notebooks': final report], Vienna: Bundesministerium für Bildung, Wissenschaft und Kultur.

Spiel, C. and Schober, B. (2003) 'Zusammenfassung des Projekts "Lebenslanges Lernen als Ziel: Welchen Beitrag kann die Schule zum Aufbau von Bildungsmotivation leisten?" [Summary of the project 'Lifelong learning as goal: what contribution does school make to build up learning motivation?'], *Erziehung und Unterricht* [Education and Teaching], 9(10): 1271–81.

Spiel, C., Lueftenegger, M., Gradinger, P. and Reimann, R. (2010) 'Zielexplikation und Standards in der Evaluationsforschung' [Explication of goals and standards of evaluation research], in H. Holling and B. Schmitz (eds) *Handbuch Statistik, Methoden und Evaluation. Reihe* [Handbook of statistics, methods and evaluation], Goettingen: Hogrefe.

Subrahmanyam, K. and Greenfield, P.M. (2008) 'Online communication and adolescent relationships', *Future Child*, 18: 119–46.

Chapter 8

The digital generation gap revisited

Constructive and dysfunctional patterns of social media usage

Jacek Pyżalski

Introduction

Looking at the usage statistics in many Western countries, we can see young people are the most frequent users of social media (e.g. Cox Communications Teen Online and Wireless Safety Survey, 2009; Diagnoza szkolna, 2009). Although benefits stemming from social media usage for psychosocial development are widely known, it is often the dark side of the internet which is believed to have an impact on young people. Young people are perceived sometimes as vulnerable victims of the internet (Buckingham, 2008). Research shows that a significant number of parents and adults are deeply concerned about the impact of the internet and cell phones on the younger generation. At times, they are convinced that excessive access to new media could dehumanize personal relations; with its equally devastating impact on young people's social lives in the communication and socialization sphere (Nissenbaum and Walker, 1998).

One of the reasons for this situation is the fact that adults' (both parents' and teachers') attitudes toward new communication technologies and the patterns of using it are different from those of young people. This may contribute to adults feeling that they and young people live 'on two different islands' where ICT usage is concerned. This situation is sometimes called the digital gap between the older generation and children and adolescents. This phenomenon is often taken as a given, and raises rather negative emotions and doubts in parents and teachers. So how can we understand the term 'generation gap'? Looking only at the frequency and length of usage (quantitative aspect) would not seem to provide a deep enough insight into the constructive and dysfunctional aspects of ICT usage among adolescents as compared with the older generation. Perhaps technical competencies, concerning the knowledge of software or internet communication channels should be considered? This would also include the social meaning of the internet and computer-mediated communication and their role in developing the social identity of users. This potential gap may also concern the perception of ICT risks such as electronic aggression or e-pornography. Depending on the meaning of a generation digital divide that is adopted for analysis, the various consequences of its existence and conclusions for media education should be discussed.

The chapter begins with the conceptual discussion concerning the definition of the digital generation divide. The potential consequences of the existence of the divide are presented and discussed, as these influence the dialogue between the generations as well as the socialization and education of the younger generation.

The empirical concept for understanding and measurement of ICT experiences and attitudes, that can be used to analyse the digital generation divide, is presented. The analysis will focus on the potentially constructive and/or dysfunctional patterns of ICT usage.

The data are presented in a way that enables an exploration of teachers' and minors' responses. Such a comparison adds a new perspective of understanding positive and dysfunctional ICT use patterns including the dynamics of interaction between professional educators and adolescents.

A closer look from this perspective contributes important information that helps us to understand the similarities and differences between teachers' and adolescents' constructive and dysfunctional ICT use patterns. Subsequently, such a perspective is vital for constructing knowledge-based programmes for media education, particularly when connected to the prevention of dysfunctional involvement in social media, e.g. engagement in cyberbullying.

The digital divide between generations: a radical view

It is unreasonable to claim that there are no differences between young people and adults when it comes to practice and beliefs concerning ICT usage, particularly with respect to the internet. There is a commonly accepted belief that young people are generally more eager to use new communication technologies and that those technologies play a vital role in their lives. Many parents and teachers do not need social science research to back up their opinion that there is a gap or divide existing between them and young people where ICT use is concerned. This digital generation gap refers to the older generation who were born in a world where the internet was not commonly used and the children born in the mid-to late 1980s and later (Herring, 2008).

Moreover, this gap is sometimes perceived as 'deeper' than the generation gaps we have experienced in the past (for example, those connected to the more traditional media, mainly television). Accordingly, young people are labelled 'digital natives', 'the digital generation', 'the net generation', 'digital citizens', 'generation i' and many similar terms (Aslanidou and Menexes, 2008; Herring, 2008; Prensky, 2001). Those terms place the young ICT users in radical opposition both to previous generations of young people as well as the generation of parents and teachers. In this context young people learn differently, communicate differently and develop differently. In some cases it is suggested that usage of new media from the early stages of the development may influence the structure of the brain of a child (Prensky 2001) – changing the character of the basic psychological processes as perception and thinking.

Such approaches explicitly or implicitly assume the existence of media determinism (Buckingham, 2008). Young people in this context are perceived as 'products' or 'victims' of the new media world that influence them in a one-directional way. Herring (2008: 76) thinks that such radical constructs 'represent contemporary youth as cyborgs, a merging of human and machine-exotic and "other"'.

Marc Prensky (2001: 1) provides the following interpretation of the consequences that stem from the generational digital divide:

> Today's students have not just changed incrementally from those of the past, nor simply changed their slang, clothes, body adornments, or styles, as has happened between generations previously. A really big discontinuity has taken place. One might even call it a 'singularity' – an event which changes things so fundamentally that there is absolutely no going back. This so-called 'singularity' is the arrival and rapid dissemination of digital technology in the last decades of the twentieth century.

It is worth noting that in his recent work, Prensky (2009) points out that due to demographic changes (teachers and parents becoming digital natives) the radical view of previous studies is no longer valid and needs reformulation. Based on this, Prensky talks about 'digital wisdom' that can improve all spheres of human life through ICT usage.

The discontinuity mentioned previously by Prensky, when understood literally, may form the basis for a real educational challenge. Prensky stated in 2001 that an educational system prepared by digital immigrants is no longer useful for digital natives. Although that statement refers to the institutional educational system, Prensky's ideas reach far beyond that context addressing the communication between the generations as such.

It is worth analysing what impact such a perspective can have on the research of young people's ICT usage. First of all, one can easily concentrate on those aspects that seem to be unique to 'digital natives'. Such dangers have been analysed by David Silverman (2006), who tended to use the term 'tourism', when describing the qualitative researchers who concentrate on finding something new while exploring the certain problem – and lose sight of anything that is similar to the phenomena they knew before. From such a perspective young people are sometimes researched in a way that is similar to researching a 'different species'. Such an approach influences the language used in the research papers and the way the research data is interpreted. For example, Aslanidou and Menexes (2008: 1376) state that 'The digital generation is being studied in many countries and efforts are being made to monitor its behaviour, and examine the use and practices of youth and their daily life at home and school.' Of course, it is not wrong when we empirically test to what extent the young internet users are different. However, at the same time we have to be very cautious when drawing conclusions from the research data we obtained. Finding, for example, a difference concerning one aspect of ICT usage is not enough to talk about revolutionary change. For example, high frequency rates of internet use or the length of daily internet usage are not sufficient indicators to draw a really thick line between the two generations since these facts do not provide any deeper insight concerning the psychosocial aspects of those differences. To emphasize this, I would like to put forward the thesis that both researchers and practitioners are often tempted to adopt the digital gap theory as a well-researched concept that does not have to be explained. This approach may be misleading as the potentially false theoretical premise (even if only partially) may easily mean a bias in the research that is based on it as well as ineffectiveness of implemented educational activities based on this theory.

Uncritical acceptance of the radical belief in the generation gap may influence everyday practices concerning broadly understood education (particularly media education) as well as the whole communication and socialization processes that take place between the two generations. This is supported by Li (2007: 378), who states:

> A technology-enhanced environment, therefore, can be viewed as a system that emerges from the interaction of its components. These components are the critical stakeholders and include students, teachers, and administrators. These stakeholder groups interact with each other and carry out certain tasks that enable the environment to function. For example, students' and teachers' beliefs about technology may affect their adoption of the tools which directly contributes to the establishment of a technology-enhanced environment.

In conclusion, the generational digital gap should not be taken for granted but rather researched as a complicated multidimensional phenomenon. This is the only way to draw conclusions from digital generation gap theory that can be theoretically and practically sound.

Moral panic and the digital gap between the generations

The digital generation gap is additionally reinforced by concerns that young people, while using technologies, may be involved in dangerous situations concerning that usage, both as victims and perpetrators. Among such situations we can list: access to inappropriate content (e.g. pornographic or violent material); risky contacts, e.g. sexual victimization or involvement in dysfunctional conduct, e.g. electronic aggression (peer aggression or cyberbullying) or internet abuse (Hasebrink *et al.*, 2009). These phenomena understandably are worrying and the carers of the children and teenagers feel that they should act in order to protect young people from the negative consequences of the ICT usage.

Questions emerge as to whether those worries will act as the basis for constructive activities or will cause a panic reaction that is not helpful in this context. There are clear signs that the second approach is quite often present among adults. As an example we can examine the situations when young people have been victims or perpetrators of various dysfunctional activities on the internet. The cases of suicides or attempted suicide of cyberbullying victims, though tragic, are often unfairly treated as examples of typical patterns of new media usage in young people. Additionally, it is often assumed as an obvious fact that cyberbullying is the worst phenomenon in terms of consequences for the victims compared to traditional bullying. Deeper analysis of the other factors (beside cyberbullying itself) that may have contributed to the decision to attempt suicide is very rarely conducted. The same refers to the analysis of the perceived impact of cyberbullying. In most cases, it is adults who state that cyberbullying is worse than traditional bullying (Pyżalski, 2010) – in very rare cases are the opinions as presented by young people themselves, analysed.

Emotional reactions, facilitated by the mass media (and social media as well), based on anecdotal situations, often lead to searches for quick, radical and very rarely good solutions to any social problems, including those stemming from ICT use among young people. Such reactions may be generally referred to as a 'moral panic'. This term was initially used by Cohen (1972: 28) who explains it in this way:

> Societies appear to be subject, every now and then, to periods of moral panic. A condition, episode, person or group of persons emerges to become defined as a threat to societal values and interests; its nature is presented in a stylized and stereotypical fashion by mass media; the moral barricades are manned by editors, bishops, politicians and other right thinking people; socially accredited experts, pronounce their diagnoses and solutions; ways of coping are evolved or (more often) resorted to; the condition then disappears, submerges or deteriorates and becomes more visible.

Shoka and Thierer (2009) emphasize that the beginning of moral panic and the New Media started with social networking and the possible contact of young people with sexual predators. Then it extended to the much wider palette of potentially dangerous aspects of internet usage.

Moral panic mixed with an oversimplified picture of the digital generation is not a very good basis for rational intergenerational cooperation in responsible media use and prevention or intervention when a dysfunctional usage occurs. As some commentators state: social scientists are not free of the influence of a moral panic (Bennett *et al.*, 2008).

The generational digital divide in the research data

This section focuses on the empirical exploration of the digital generation gap, and concerns both constructive and dysfunctional social media usage patterns. The data will be restricted to two groups: students and teachers.

The digital generation gap framework concept, adopted for this study, covers the following aspects:

1. general attitudes concerning the impact of new communication technologies;
2. internet competencies: the positive patterns of internet use and technical competencies;
3. dysfunctional internet usage: internet abuse, or using the internet as a way to escape from the offline world;
4. attitudes toward electronic aggression.

The presented concept has been measured across two population studies prepared and conducted by the author and his team, namely:

1. a representative survey of a sample of 14–15-year-old Polish adolescents (N = 2143) (Cyberbullying as a new peer aggression type);
2. electronic aggression as a new problem of teachers' occupational health: a survey of 600 Polish teachers.

Data for both projects was gathered in the same year (2009) which provide the unique opportunity to use them for a comparison and explore more deeply the construct known as the digital generation gap.

General attitudes concerning the impact of new communication technologies

Young people as a whole are often perceived as internet-savvy and totally optimistic about new communication technologies and social media. In order to empirically check this assumption, questions concerning the impact of new communication technologies were adapted from the population survey conducted by CBOS (2008). The respondents in that study were asked whether the world with new communication technologies (the internet, mobile phones) is becoming better or worse. That question was then repeated in the questionnaire for teachers and students as: 'Is the world with the new technologies like the Internet or mobile phones becoming better or worse?' From all three groups, as expected, the adolescents were the most optimistic. Some 58 per cent of them think that the world is a better place with new technologies. At the same time almost 14.5 per cent think that it is neither better nor worse, and 22.5 per cent cannot assess it. Only 5 per cent think that the world is worse with new communication technologies. What is interesting is that teachers were more pessimistic than students but also more pessimistic than the general Polish population with the respective figures: 26 per cent (better), 36 per cent (neither better nor worse), 27 per cent (cannot assess it) and 16 per cent (worse).

Social media competencies: positive patterns of internet usage and technical competences

Both teachers and students completed the ICT competence and practice scale. This scale covers a subjective assessment of respondents' ICT skills as well as the facts concerning their practices concerning internet usage.

When it comes to the competence of creating a webpage – less than a half of students report that they can do it without external help (46.9 per cent). This competence is reported by more than a quarter of the teachers. This results demonstrates that young people may not be as cyber-savvy as might be expected – but of course only when this simple competence indicator is considered. When we compare the difference between the students and the teachers, the gap also seems to be narrower than may be initially expected by popular beliefs, underlying the total incongruence between both groups in terms of internet competence (Herring, 2008).

Teachers and students similarly assessed their ICT skills in relation to people from their social groups. A little less than one-third of the respondents in both groups claim that their personal skills are higher than those of colleagues and friends (Table 8.1). Of course while interpreting this, we should not forget that respondents subjectively assess their competences and the reference groups are different in both cases. Answers to those questions can by no means be understood as indicators of objective ICT competencies.

Table 8.1 Teachers' and students' answers on the scale of ICT competence and practice

	Teachers (N=600) (%)	*Students (N=2143) (%)*
I can create a webpage independently	26.5	46.9
I can operate a computer better than most of my friends	34	31.9
I am an expert when it comes to computer usage	5.3	13.9
I publish on the Net the content I produce (e.g. graphics, music I compose)	6.4	30.3
I run my own webpage or blog	5.8	25.9

Almost 14 per cent of students claim to be experts in ICT use, compared to only 5 per cent of the teachers.

Special focus should be placed on the particular group of internet users called 'content creators'. According to Lenhart and Madden (2005: 3), content creators are the users who actively produce and publish materials online. According to these authors' study in the year 2005, approximately 57 per cent of American teenagers reported having done one or more of the following activities: create a blog; create or work on a personal webpage; create or work on a webpage for school, a friend, or an organization; share original content such as artwork, photos, stories, or videos online; or remix content found online into a new creation. Particularly active in those activities were young bloggers, especially girls. In our study this concept has been operationalized by two items, namely 'I publish on the Net the content I produce (e.g. graphics, music I compose)' and 'I run my own webpage or blog'. In both cases young people are more active than teachers. Almost five times more young people than teachers publish on the Net the content they produce and three times more run their own webpage or blog. This difference should be perceived as crucial. Many of today's adults seem to use the new media in a mode similar to how they use traditional ones. They are rather passive consumers not active media creators. The opposite situation is observed in young people – many of them use the net as the potential 'stage' to present the content they produce to narrow or wider audiences. Of course, we cannot forget that this concerns only a certain proportion of the respondents – only 14.3 per cent reported being involved in the two 'content-creating' activities. At the same time there are both young people who are content creators and others who are only internet content consumers (ibid.). Of course such consumers can be found among adults in even higher proportions but here the research shows differences. For example, Lenhart and Madden found that young people download videos twice as often as adults.

Dysfunctional internet usage: internet abuse, using the internet as a way to escape from the offline world

Teachers and students also completed the Dysfunctional Internet Usage scale. This scale covers some aspects of internet abuse (Morhan-Martin, 2008) as well as treating the internet as a way to escape from offline life.

Dysfunctional internet usage indicators are more prevalent among young people. They are much more likely than teachers to be 'immersed' in ICT. Some 37.8 per cent of young respondents are bored if they have no connection to the internet even for one day (Table 8.2). One in five spends the whole day in front of computer whenever they have a free day. Almost one in three students has a lot of secrets concerning their online activities. All those things are very rarely reported by teachers as their personal experience.

Table 8.2 Teachers' and students' answers on the scale Dysfunctional Internet Usage

	Teachers N = 600 (%)	*Students* (N = 2143) (%)
I notice I spend more and more time in front of the computer	27	45.7
I am bored when I have no connection to the internet for even one day	8	37.8
When I have got a free day I spend the whole time in front of a computer	3.7	19.1
It is better no one knows what I do on the internet	5.1	32.7
I have got a lot of secrets concerning what I do while surfing the internet	3.4	32.2
I often skip sleep at nights because I use the internet	5.6	12.3
I feel better in the virtual world than in the real one	1.9	15.6
Among all my interests the internet is first place	5.6	10.3

It is worth reporting that particular personal significance of the internet is reported by moderate rates of the students. Only 10.3 per cent claim the internet is the most important of all their interests and 15.6 feel better in the online world in comparison to the offline one. This again is against popular belief and the common-sense knowledge concerning the involvement of young people in new communication technologies, particularly the internet.

Attitudes of teachers and students toward electronic aggression

Both teachers and students presented their perceptions of the consequences of cyberbullying (Smith, *et al.*, 2008; Dooley *et al.*, 2009) and traditional face-to-face peer aggression. Almost 48 per cent of the teachers think that cyberbullying consequences are worse in comparison to consequences of traditional bullying. This perception is shared only by 13.1 per cent of students. The totally opposite proportion concerns the belief that traditional bullying is worse than cyberbullying. Such opinion is backed by 39.3 per cent of the students and only 2.8 of the teachers. This example clearly shows that attitudes concerning the dangers connected to social media may be totally different in adults and young people.

Conclusion

The data presented seem to back up the interpretation provided by Herring (2008). This author states that 'the so-called "Internet Generation" is in fact a transitional generation, in which young Internet users are characterized to varying degrees by a dual consciousness of both their own and an adult perspective' (ibid.: 72). In some respects the differences between teachers and students have been less than expected. For example, those who think that they can better operate ICTs in comparison to their friends have been similar in both groups. However, there were areas where the differences were vital. At least two aspects should be noted there. First, young people are more often content creators, who actively produce and publish their internet content (Lenhart and Madden, 2005). Second, young people more often than teachers consider cyberbullying to be a less serious peer aggression type as compared to face-to-face bullying.

The general conclusion of all the data concerning the aspects explored may be that deep involvement in cyberactivities (both positive and dysfunctional) concerns only the minority of young people. Taking this into account we also have to remember that an oversimplified picture of today's younger generation as the 'cyber'group, whose activities are all focused only on ICTs and online life, may be far from true. Of course, that does not mean that we should ignore differences where we really find them.

The research methodology and scope could thus really benefit from a paradigm shift from the one that is fascinated 'with *technologies*, to a focus on young *people* themselves and their communicative needs as they happen to be expressed through particular media' (Herring, 2008: 72). This means we should focus on deeper research of particular aspects of new media use in young people. Such aspects are not always obvious and require much deeper investigation. This knowledge may facilitate constructive communication between generations, particularly within educational settings.

Acknowledgements

The author received grants for both surveys: 'Cyberbullying as a new peer aggression type (Grant No. 106 067735) – a qualitative and quantitative research study of a sample of 2143 Polish adolescents (14–15 y.o.)'. This grant is affiliated to Wyższa Szkoła Pedagogiczna w Łodzi. 'Electronic aggression as a new problem of teachers' occupational health' (Grant No. IMP 8.5) – a survey of 600 teachers. This grant is affiliated to the Nofer Institute of Occupational Medicine.

References

Aslanidou, S.and Menexes, G. (2008) 'Youth and the Internet: uses and practices in the home', *Computers & Education*, 51: 1375–91.

Bennett, S., Maton, K. and Kervin, L. (2008) 'The "digital natives" debate: a critical review of the evidence', *British Journal of Educational Technology*, 39(5): 775–86.

Buckingham, D. (2008) 'Nowe media – nowe postaci dzieciństwa? Zmieniające się środowisko kulturowe dzieci w erze technologii cyfrowej', in M.J. Kehilly (ed.) *Wprowadzenie do badań nad dzieciństwem*, Kraków: WAM.

CBOS (2008) *Polacy w Sieci*, available at: BS/58/2008.http://www.cbos.pl/SPISKOM.POL/2008/K_058_08.PDF (accessed 10 May 2010).

Cohen, S. (1972) *Folk Devils and Moral Panics*, Harmondsworth: Penguin.

Cox Communications Teen Online & Wireless Safety Survey (2009) *Teen Online & Wireless Safety Survey: Cyberbullying, Sexting and Parental Control*, in Partnership with National Center for Missing and Exploited Children (NCMEC) and John Walsh.

Diagnoza szkolna (2009) 'Raport roczny programu społecznego "Szkoła bez przemocy"', available at: http://www.szkolabezprzemocy.pl/938,raport-roczny-2009 (accessed 10 May 2010).

Dooley, J.J., Pyżalski, J. and Cross, D. (2009) 'Cyberbullying versus face-to-face bullying: a theoretical and conceptual review', *Zeitschrift für Psychologie* [Journal of Psychology], 217: 182–8.

Hasebrink, U., Livingstone, S., Haddon, L. and Ólafsson, K. (2009) *Comparing Children's Online Opportunities and Risks across Europe: Cross-national Comparisons for EU Kids Online*, 2nd edn. London: London School of Economics and Political Science, Departmentof Media and Communications, available at: http://eprints.lse.ac.uk24368/1/D3.2_Report-Cross_-national_comparisons-2nd edition.pdf / (accessed 10 November 2010).

Herring, S.C. (2008) 'Questioning the generational divide: technological exoticism and adult constructions of online youth identity', in D. Buckingham (ed.) *Youth Identity and the Digital Media*, Cambridge, MA: MIT Press.

Ito, M., Horst, H. A., Bittanti, M., boyd, D., Stephenson, B. H., Lange, P. G., Pascoe, C. J. and Robinson, L. (2008) *Living and Learning with New Media: Summary of Findings from the Digital Youth Project*, The John D. and Catherine T. MacArthur Foundation Reports on Digital Media and Learning, Cambridge, MA: MIT Press.

Lenhart, A. and Madden, M. (2005) *Teen Content Creators and Consumers*, available at: http:// www.pewinternet.org/Reports/2005/Teen-Content-Creators-and-Consumers/3-Teens-as-Content-Creators/02-More-than-half-of-online-teens-are-Content-Creators.aspx?r=1 (accessed 10 May 2010).

Li, Q. (2007) 'Student and teacher views about technology: a tale of two cities?' *Journal of Research on Technology in Education*, 39(4): 377–97.

Morhan-Martin, J. (2008) 'Internet abuse: emerging trends and lingering questions', in A. Barak (ed.) *Psychological Aspects of Cyberspace: Theory, Research, Applications*, Cambridge: Cambridge University Press.

Nissenbaum, H. and Walker, D. (1998) 'Will computers dehumanize education? A grounded approach to values at risk', *Technology in Society*, 20(3): 237–73.

Prensky, M. (2001) 'Digital natives, digital immigrants', Parts I and II, *ON the Horizon*, 9(5).

——(2009) 'H. sapiens digital: from digital immigrants and digital natives to digital wisdom', available at: http://www.innovateonline.info/pdf/vol5_issue3/H._Sapiens_Digital–from_Digital_Immigrants_and_Digital_Natives_to_Digital_Wisdom.pdf (accessed 15 December 2010).

Pyżalski J. (2010) 'Polscy nauczyciele i uczniowie a agresja elektroniczna – zarys teoretyczny i najnowsze wyniki bada& nacute;', in M. Jędrzejko and D. Sarzała (eds) *Człowiek i uzależnienia, Pułtusk-Warszawa: Akademia Humanistyczna im.* Aleksandra Gieysztora, Oficyna Wydawnicza ASPRA-JR.

Shoka, B. and Thierer, A. (2009) 'Cyberbullying legislation: why education is preferable to regulation', *Progress on Point*, 16: 1–26.

Silverman, D. (2006) *Interpreting Qualitative Data*. London: Sage.

Smith, P.K., Mahdavi, J., Carvalho, M., Fisher, S., Russell, S., and Tippett, N. (2008) 'Cyberbullying, its forms and impact in secondary school pupils', *Journal of Child Psychology and Psychiatry*, 49: 376–85.

Chapter 9

Kids in the fast lane

Achieving well-being through online support

Zack Hayat and Yair Amichai-Hamburger

Introduction

The Internet today is part of the landscape of our daily lives. For young people, the Internet is a constant companion, both at school and at home, computers are now found in kinder-gardens all over the world. At school, lessons take place using the Internet and students are expected to make full use of it as a research tool to do their homework. At home the Internet is also the leisure tool of choice for many children and young people, and it is through the Internet that they relax and organize their social lives. The Internet has changed the way we engage socially, commercially and educationally and is certainly here to stay. While there are many positive aspects to its use, the rapid growth of the Internet has largely outpaced our ability to understand the effects of this dynamic medium, both on our society in general and on people individually (Amichai-Hamburger and Barak, 2009). When we try to bridge the gap and understand the reach of the Internet, research has tended to focus on the experience of adult users (e.g. Mok *et al.*, 2010) while the experience of young people and children is less well documented.

The aim of this chapter is to review the literature dealing with the relationship between Internet use and the psychological and psychosocial well-being of children between the ages of 9 and 17, otherwise known as the 'N-Genners' (Net Generation). The Internet has become a 'necessity' for N-Genners, but the drawbacks of sustained Internet use are still unclear. This chapter discusses current trends in Internet use among children and teenagers, with a focus on the more troublesome aspects of such usage: aspects that might require responsible adults to pay more attention to how this medium is being used today by children and teenagers. While exploring these troublesome aspects, we also discuss how our knowledge about the potential negative implications of Internet usage can, in turn, help us in thinking of ways to productively deal with these troublesome issues. With the publication of such findings, the public has an opportunity to become more informed and educated as to the advantages and dangers of young people's Internet use. This knowledge, will, it is envisioned, allow those responsible adults, including parents and teachers to make informed and appropriate decisions which will facilitate effective use of the Internet by young people. In addition, creators and designers of technology will, we trust, use such knowledge to produce technology that is psychologically healthy and socially adaptive.

Internet usage

The starting age for Internet users is steadily decreasing while the average amount of time spent on the Internet is rapidly increasing (Mok *et al.*, 2010). According to the Current US Population Survey: School Enrollment and Computer Use Supplement (SECS), about two-thirds of children in nursery school, and 80 percent of kindergartners use computers, indicating that computer use among the very young is well established.

The proportion of children using computers continues to increase with age, such that by the time children reach high school, almost 97 percent of students in grades 9–12 are active computer users. In terms of Internet use, 23 percent of children in nursery school use the Internet, and this number rises to 50 percent by grade 3 and to 79 percent in grades 9–12 (Norris, 2010). In other words, the Internet is becoming a larger factor in the lives of people at progressively younger ages. Thus, the question of how Internet usage effects children and teenagers is becoming increasingly pertinent.

Teens represent a unique audience in that their online usage patterns are distinctly different from those of other age groups. When looking at the major activities of the N-Genners online, we learn that 73 percent of wired American teens now use social networking websites, as compared to 47 percent of online adults. 62 percent of online teens access news about current events and politics online, and 48 percent of wired teens have purchased items online, for example, books, clothing or music (Lenhart *et al.*, 2010). According to Livingstone (2003), typical Internet usage of young children can be described in the following three dimensions: (1) entertainment; (2) education; and (3) edutainment. With that said, it is more difficult to talk about children's Internet usage patterns as the vast majority of the studies looking at this aspect are focused on media usage in general (e.g. TV vs. Internet). These studies pay little attention to the online usage patterns of children (Genevieve, 2010).

It is evident that children and young people are going to be influenced by the extent of their use of the Internet, socially and psychologically. What those impacts are, and how we can learn from them to find ways of providing positive experiences for relationships in educational settings is the challenge.

Effects of Internet use on children and teenagers

The following sections discuss relevant research and theory, which is largely speculative at this time, which strives to explain the effects of Internet use on children and teenagers. Based on the relevant literature, seven main areas of impact have been identified: (1) parent–child interaction; (2) cyberbulling; (3) the notion of play; (4) the development of social causal reasoning skills; (5) vulnerability to social isolation; (6) Internet use and critical periods in development; and (7) the effects of multitasking.

Parent–child interaction

Parent–child interaction indubitably plays an important role in the social and emotional development of children (Sefton-Green and Buckingham, 1996). Parents often impart moral guidance, social skills modeling, and emotional lessons to their children (Crain, 2000). Yet, research indicates that contemporary children are not spending enough time with their parents (Lynda and Kerry, 2010). Basing her assessment on case studies and observations in homes and classrooms around the United States, developmental and educational psychologist Jane Healy (2004) hypothesizes that children's Internet use may detract significantly from parent–child interactions, if it occurs at times when parents and children would normally interact. If time spent on the Internet, either by parents or children, comes between parents and their children during the formative years, then the child's emotional and social development might be impacted.

Another issue that the Internet and computers in general raises for parents with Net generation children is that the parents are unaware of just what their children do during the hours they spend on the computer and online (Subrahmanyam and Lin, 2007). Surveys indicate that most parents believe that the computer/Internet is an important academic tool, and that their children are reaping great benefits from their time with the machine (Heim *et al.*, 2007). Based on case studies and observations, Healy suggests that these reasons might explain why parents often find it difficult to set time limits on this activity, whereas they might find it easier to set limits on their child's TV watching, which is more of a known quantity (Healy, 2004). This situation can subvert the traditional parental authority paradigm, leading to problematic family relations, and hindering the development of the child's social skills (ibid.).

This lack of sufficiant guidance and supervision regarding children's Internet usage begs the question as to the potential dangers that such usage might entail. One of those dangers is exposure to pornography and other unsuitable material (e.g. graphic images of any kind that are inappropriate for children of various ages).

A small, but consistent body of research from various parts of the world has shown that a considerable number of adolescents have contact with sexually explicit Internet material. Whether intentionally or not, 42 percent of children in the age group of 8–16 years have viewed sexually explicit Internet material, and children are first exposed to such content at the age of

11 years on average. The largest consumers of sexually explicit Internet material are from the 12–17-year-old age group (Wolak *et al.*, 2007). Websites containing sexually explicit, violent or otherwise inappropriate material are readily available to children who use the Internet without parental supervision. Healy makes a comparison between allowing a child to wander freely and unsupervised on the Internet and 'dropping your child off alone in a bad neighborhood' (Healy, 2004, p. 259).

These issues introduce the critical need for parental supervision (Wang *et al.*, 2009). Lwin, Stanaland and Miyazaki (2008) distinguish between two types of strategies: restrictive mediation and active mediation. Restrictive mediation implies the definition of rules about Internet usage. These rules can be related to timing or to the use of or access to specific Internet applications. Restrictive mediation precedes actual Internet usage. Active mediation is linked to actual Internet usage. Parents adopt an active position and stay around, talk about Internet usage, or check the computer screen.

Some authors question the impact of active or restrictive mediation (Lee and Chae, 2007). However, the study by Lwin *et al.* (2008) shows that both active and restrictive mediation help to lower the chance of risky online contact (for example, passing personal information). They therefore prefer to label the combination of active and restrictive mediation 'selective mediation'. They also found that 'promotive mediation' is effective when parents especially adopt active mediation and to a lesser extent restrictive mediation. The least effective is a *laissez-faire* style; they found that active and restrictive mediation predicted a lower level of unsafe Internet behavior.

Cyberbulling

Another impact we should be aware of lies with cyberbullying. Cyberbullying and/or cyberharassment occur when someone uses electronic media such as a cell phone or the Internet technology to embarrass, shame, or humiliate another person (Patchin and Hinduja, 2006).

Cyberbullying was recently blamed for the suicide of a student at an American university who was 'outed' through the Internet. Ybarra, Mitchell, Wolak, and Finkelhor (2006) reported that approximately 9 in 10 children and adolescents report using the Internet and that 9 percent of respondents to the Second Youth Internet Safety Survey reported being the target of cyberharassment at least once in the previous year, a 50 percent increase from the First Internet Safety Study (Finkelhor *et al.*, 2008) conducted just five years earlier.

In recent years, several cases have come to the public's attention in which individuals have harassed, stalked and killed unsuspecting children whom they have met online. Internet users, particularly young users, may be unaware of the dangers of revealing personal information (address, phone number, school location, etc.) to the total strangers they meet online.

Cyberbullying is becoming an increasingly serious issue, which requires awareness by school officials, staff, and parents to effectively address and contain. Ybarra *et al.* (2006) suggested that as a result of the significant increase in the report of cyberharassment from 2000 to 2005, vigilance is still called for on the part of those who interact with young people.

Patchin and Hinduja (2006) noted a lack of supervision in cyberspace. They pointed out that although chat rooms are periodically monitored in an attempt to regulate conversation and eliminate offensive individuals, personal messages are viewable only by the sender and recipient. Further, e-mail or text messages sent via computer or cellular phone are not subject to any type of monitoring or censorship. In addition, an increasingly large number of adolescents have access to computers in the privacy of their bedrooms and in many cases the adolescent knows more about the use of computers and cellular phones than his or her parents and is, therefore, able to operate without fear of being discovered regarding their involvement in cyberbullying or being a victim.

Juvonen and Gross (2008) looked at prevention tactics used by their respondents to protect themselves from cyberbullying. The most common tactic used was blocking a particular screen name, which 67 percent of the respondents reported they had done in the past. One-third reported having restricted a particular name from their buddy list, while 26 percent had switched their screen name and a similar number had sent a warning to the perpetrator to prevent cyberbullying. Juvonen and Gross reported that these tactics are particularly relevant to Instant Messaging, which was reported as one of the most prevalent forms of electronic communication by their sample.

One assumption supported by the data collected by Juvonen and Gross was that youth would not report incidents of cyberbullying to adults which was in contrast to Ybarra *et al.* who in 2006 reported that two in three victims of cyberbullying reported that they had disclosed the incident to a parent or other person in authority. The response found by Juvonen and Gross was, in fact, overwhelming, as 90 percent of their respondents reported not telling adults about incidents of cyberbullying. The most common reasons reported for not telling were a stated belief by 50 percent of the respondents that they needed to learn to deal with it by themselves and a fear, reported by 31 percent, that if their parents found out, their access to the Internet would be restricted. This concern was even more pronounced for younger respondents, particularly girls from age 12–14.

The direction that looks most promising in dealing with cyberbullying comes from the Rodkin and Hodges (2003) study. Rodkin and Hodges found that those children who can develop positive peer relations with a wide group of people are to a great extent protected from victimization. Rodkin and Hodges argue that when victimized children are appropriately integrated into their peer ecologies, forming the right friendships and heading off destructive antipathies, present and future risks can be reduced.

Therefore, a potentially powerful intervention for potential cyberbullying victims is training to enhance social networking techniques as well as efforts to create a more socially accepting environment at school through teacher modeling and encouragement. They also suggested that it would be beneficial for school psychologists to develop an anti-bullying curriculum, which will both enhance the understanding of current research as well as support efforts to reduce children's aggression towards one another. More research needs to be conducted in this area in order to improve anti-bullying measures that involve educators and parents, as all students are potentially at risk.

Play

Nie and Erbring (2000) and other experts argue that time spent using the Internet leaves less time for face-to-face and other traditional means of social interaction, which may lead to social isolation and other negative social and emotional consequences. It is well known that play and peer interaction in general form an important part of a healthy childhood, and that a child's spontaneous, natural play behaviors promote healthy physical, social and psychological development (Azar, 2002). Some researchers are concerned that Internet use may displace a child's playtime or time for face-to-face social interaction with peers, and therefore interfere with healthy physical, social and emotional development (Subrahmanyam and Lin, 2007).

Play, formal or informal, solitary or in company, provides a child with regular exercise (Azar, 2002). It is empirically proven and generally accepted that exercise provides health and psychological benefits for people of all ages (King *et al.*, 2002). Physical fitness, attained through regular exercise, also enhances secretion of important neurochemicals, such as endorphins. Endorphins, short for 'endogenous morphines' are secreted by the pituitary gland during exercise. Endorphins elevate mood, stave off stress and depression, and inhibit physical pain (Dewey and Steinberg, 2001). Other cognitive and psychological sequelae of physical fitness include improved mental functions such as learning and memory, and improved self-confidence (Sapolsky, 1998). Internet use is an inherently sedentary and mostly solitary activity. In fact, more than 60 percent of all computer time is spent alone (Norris, 2010). If the hours a child spends using the Internet replace some of the time he or she might otherwise spend engaging in active play, then the Internet might deprive the child of some of the health benefits which play naturally provides (Healy, 2004).

In addition, play can help children deal with natural aggressive instincts, modulate affect, and develop a sense of self (Azar, 2002). Playing with parents or other adults also provides the child with useful tools for emotional and social development, including adult role modeling and mentorship. Play with adults also introduces the child to adult social interactions and expectations, which become more important as the child grows older.

Some research even suggests that depriving a child of play may interfere with brain maturation, and may increase the likelihood of impulsivity and poor self-control due to a disruption in frontal lobe development (ibid.). It is clear that play has an important role in the physical, psychological and social development of children. Some researchers are concerned that if online time displaces a child's 'physical' playtime or time for social interactions, then it is possible that the development of healthy social skills will be disturbed (Tarpley, 2002). More research must be done in order to form a clearer picture of the time displacement effects of Internet use on children and teenagers.

It is important to note that physical inactivity or lack of social interactions among children and youth remains a significant issue that will likely only be solved through a multifaceted approach that includes education and structured interventions combined with the provision of enticing opportunities for different activities. The new generation of both online and offline Active Video Games (AVGs) are an emerging technology that has the potential to address the latter. Preliminary evidence seems to support AVG play as an enjoyable medium for self-directed physical activity of light to moderate intensity (Leatherdale *et al.*, 2010), as well as offering a socially interactive playtime (Lwin *et al.*, 2008). It remains to be seen whether AVGs can be used effectively in the long term to help motivate increased daily physical activity and decreased sedentary pastimes, as well as the way in which they encourage social interactivity.

Development of social causal reasoning skills

Social causal reasoning skills are 'the ability to infer how someone else might be feeling' (Healy, 1998, p. 212). Research indicates that this important skill set usually begins to emerge in children as early as age 3 (Frye *et al.*, 1995), and continues to develop through everyday human interactions (Hickling and Wellman, 2001), in which the child learns to recognize the cause and effect relationship between his or her words or actions and the responses of others. Both children and adults learn to use experiences in the past to predict the outcome of new social situations, and to modify their behavior in accordance with the situation and desired outcome (Brown *et al.*, 1989).

In addition to routine social encounters, play provides another context in which children can learn these skills (Azar, 2002). Developmental and educational psychologist Jane Healy suggests that when parents allow cyberworlds to replace face-to-face social encounters, children's social skill development may be at risk (Healy, 2004). In particular, Healy suggests that time spent online may interfere with or displace a child's opportunity to develop social causal reasoning skills through social interaction and play. If a child does not acquire adequate skills for interpersonal functioning, he or she is more likely to face the pain of social rejection, and to develop emotional problems later (Wodarski and Hedrick, 1987).

It is important to note that communication technologies that were popular among adolescents in the 1990s, such as MUDs (Multi-User Dungeons) and public chat rooms, were typically used for communication between strangers. However, in recent years, several communication technologies, such as Instant Messaging (IM) and social networking sites like Facebook, have been developed which encourage adolescents to communicate with existing friends. European and US studies have shown that 84 percent (e.g., Gross, 2004) to 88 percent (e.g., Valkenburg and Peter, 2007) of adolescents use IM for communication with existing friends. These patterns suggest that rather than hindering opportunities for social interaction, Internet usage can actually supplement existing social relationships with peers. By doing so, the Internet can actually offer an additional channel for promoting one's social causal reasoning skills.

Vulnerability to social isolation

Research suggests that a sense of 'social connectedness' is very important for healthy adolescent development, and that social experimentation is a natural phenomenon in this developmental stage (Newman and Newman, 2001). Social isolation or lack of group membership may increase an adolescent's chance of encountering depression, lowered self-esteem, anti-social tendencies, and high-risk behaviors later in life (Sylwester, 1997). Some research points to a potential link between adolescent social behaviors and Internet use, which may signal a heightened vulnerability in this age range to social isolation and loneliness in connection with Internet use (Subrahmanyam and Lin, 2007).

In addition, researchers hypothesize that teenagers may experience greater loneliness and depression than adults because adolescents spend far more online time than adults. This impact relates to their spending more time corresponding with weak tie relationships, those people with whom they had no relationship other than the Internet-based correspondence, and less time communicating with strong ties, than adults did (ibid.).

Shelley Turkle, an MIT psychologist and pioneering researcher in the area of human–computer interaction, has analysed the interface between online social worlds and human identity (Turkle *et al.*, 2006). Of interest here is Turkle's view of the social and psychological effects of teenagers' participation in online communities. Based on an extensive series of interviews, Turkle expresses ambivalence regarding the social potential of online communities for teenagers. On the one hand, she believes that virtual social environments may allow teens to experiment with their identities in safe and positive ways, and to explore various types of social discourse in a relatively non-threatening environment. On the other hand, this type of social activity may break down important boundaries between virtual and real life, causing teenagers to lose sight of the importance of real emotion and traditionally established relationships (ibid.).

Additional research is needed to better understand the impact of online social venues on adolescents. One important trend that has significant importance within this context is that based on current studies, the vast majority of adolescents in Western countries have access to the Internet (e.g., Lenhart and Madden, 2007). At such high access rates, a negative effect of the Internet on social connectedness is less likely because adolescents have more opportunities to maintain their social network through this medium.

Internet use and critical periods in development

Perhaps the most fundamental and far-reaching set of ideas regarding the effects of Internet use on children is based on the idea that use of technology interferes with a child's neurological and psychological development during critical periods (Healy, 2004). This area of investigation is fairly new, and much more research is needed to build a substantive body of literature based on empirical evidence.

A critical period is 'a time in early development during which some event, such as an experience or presence of a hormone, has a long-lasting effect' (Kalat, 1998, p. 523). The term 'critical period' traditionally refers to phases in neurological and other physiological development during which environmental events and experience have a long-term impact on the development of synaptic connections and neurochemical patterns. However, the term 'critical period' has also been used to describe periods of intense vulnerability during emotional and psychological development, during which an event or disturbance may have unusually far-reaching effects on development, including changes in brain function (Helmeke *et al.*, 2001). Since children and adolescents are still developing both physically and psychologically, their interaction and experience with technology (i.e. computers, software, and the Internet) have the potential to impact them profoundly during critical periods (Healy, 2004). It has been argued that if a young person spends significant amounts of time playing computer games: software, video games, Internet or other video game programs, and those games fail to stimulate still-developing brain areas, it is possible that this individual will suffer from poor frontal lobe development and sequelae. Specifically, researchers are concerned that stunted frontal lobe development will yield children who have concentration problems, difficulty modulating emotions and controlling impulses, particularly anti-social or unethical impulses, and an increased probability of violent behavior in later life (Doğan, 2006).

On an experiential level, many factors suggest that excessive Internet use may predispose a young person to develop impulsive behavior. The example of a child playing a computer game illustrates this point. When a child plays a game, his/her goal may be to blow up a bad guy, get through a maze, avoid being shot and reach a prize. As discussed previously, an individual's physiological reactions to virtual events mirror the body's reaction to real world events. However, unlike a real situation, the child can end or restart a game with the click of a button at any moment he chooses, with no real-life consequences. Similarly, if a child or adolescent is engaging in a virtual conversation with a stranger on the Internet, and wants to end the conversation, he/she can do so by pushing a button, without consequences.

Turkle *et al.* (2006) noted that some children have difficulty making a distinction or understanding boundaries between real and virtual life while playing video games. Given that some children do have trouble distinguishing real life from computer-simulated life, a habit of disregarding consequences established through repeated experiences in the virtual world might translate into a similar pattern of disregard for consequences in the real world. Therefore, the child who plays video games excessively may begin to display impulsive behavior in the real world.

On a similar note, some experts suggest that heavy Internet use in the developing child may lead to more impulsive and less effective approaches to problem solving, management of information, and goal setting. Gavriel Salomon at the Hebrew University of Jerusalem (as cited in Healy, 2004) reports that visual media is a relatively concrete way to present information, and thus requires less 'mental elaboration' than extracting meaning from written text. He argues that this manner of information presentation may lead to superficial processing, and decreased perseverance.

On the Internet, instant gratification is easily achieved, while using impulsive trial and error guesswork to solve problems or seek information. Though it is difficult to quantify or assess empirically, some suggest that the Internet rewards this type of problem solving at the expense of critical and methodical thinking (Healy, 2004). Some psychologists, educators and parents are troubled that this chaotic approach to problem solving and information processing is becoming a poor replacement for applying analytical and synthetic thought to material, making connections, and generating meaningful and creative ideas. There is a concern that the line has become blurred between knowledge extraction and information gathering among members of the Net generation (ibid.). Healy expresses this concern: 'Don't stop to think, don't work the problem through … just jump in and try something – if it doesn't work you can blow it up, start again or switch programs' (ibid., p. 46).

The Internet represents both risks and opportunities for young people at *critical periods in development*. To better support youth in critical periods in their development, we need more research to understand which youth may be most susceptible and to develop targeted interventions to protect them. The Internet also has many positive aspects and can be used to enhance youth learning and empowerment. The Internet can also serve as a tremendous health resource and can be used to cheaply deliver interventions; we need to understand how to better implement them to enhance their effectiveness especially in critical periods in youth development.

Is multi-tasking working at all?

Another important aspect regarding N-Genners' Internet usage is their 'ever-growing need to multitask' (Fox *et al*., 2008). Parents often describe how their teenage child performs their at-home tasks while performing other tasks (e.g. listening to music on their portable digital music player or checking their Facebook pages). Some researchers describe the situation at home for today's youth as media 'saturation' through technology (Roberts *et al*., 2005). Research examining the behavior of members of the youngest generation suggests that they are multitasking more frequently when compared to non N-Genners (Carrier *et al*., 2009).

Veen and Vrakking (2006) offered the term 'Homo Zappiens' to single out the new young generations from the ones before. They suggest that this new generation is characterized among others by an integrative approach to technology, part of which is the high ability to multitask.

However, the whole issue of multitasking efficiency generally speaking, and particularly for children, is complex. Medina (2009), a molecular biologist, argues that multitasking in terms of being able to pay attention to several tasks simultaneously is no more than a myth. Our brain can only focus on one thing at a time. It is only capable of sequential processing. What appears to be multitasking is actually a task gagging. The whole idea of effective multitasking seems to run contrary to the concept of *flow* which stands for a mental state in which a person in an activity is fully immersed in a feeling of that activity. It is a single-minded immersion. It is argued that the flow experiences in life are actually the peak experiences and are a great source of well-being (Asakawa, 2010). Recent studies have demonstrated that multitasking does not work. It seems that juggling tasks leads to greater inefficiency in performing each individual task, namely that more mistakes are made, and it takes significantly longer as compared to sequential work (Ophira *et al*., 2009; Fox *et al*., 2008; Bowman *et al*., 2009). If multitasking characterizes young people's lives, its impact on affectivity is doubtful, and its impact on well-being is unclear, we need much more research in this field to understand the real effects of multitasking (Bennett *et al*, 2008).

Conclusion

When looking at current Internet usage patterns, we witness a major increase in Internet use, especially among children between the ages of 9 and 17. Review of the literature yields some empirical information and much speculation and controversy regarding the psychological and social impact of Internet use on children and teenagers. It is clear from the review presented in this chapter that alongside the benefits that might be gained from Internet usage, there are some dangers facing today's youth that we as a society need to be aware of. In light of the current Internet usage trends, and the predictions regarding its future use, it is clear that research must continue to address empirically the ways in which the Internet is affecting children growing up in the ICT age. Creators and consumers of technology and makers of public policy must be informed about the impact of the Internet on humans, and particularly on the most rapidly expanding online population, children and adolescents. Ideally, creators of new technology, educators and parents will consider the literature presented in this chapter, in order to produce, and use, technologies in a way that provide real benefit and minimal danger.

References

Amichai-Hamburger, Y. and Barak, A. (2009) 'Internet and well-being', in Y. Amichai-Hamburger (ed.) *Technology and Well-being*, Cambridge: Cambridge University Press, pp. 34–76.

Asakawa, K. (2010) 'Flow experience, culture, and well-being: how do autotelic Japanese college students feel, behave, and think in their daily lives?', *Journal of Happiness Studies*, 11(2): 205–23.

Azar, B. (2002) 'It's more than fun and games', *APA Monitor*, available at: http://www.apa.org/monitor/mar02/morefun.aspx (accessed 29 October 2010).

Bennett, S., Maton, K. and Kervin, L. (2008) 'The "digital natives" debate: a critical review of the evidence', *British Journal of Educational Technology*, 39(5): 775–86.

Bowman, L.L., Levine, L.E., Waite, B.M. and Gendron, M. (2009) 'Can students really multitask? An experimental study of instant messaging while reading', *Computers & Education*, 54(4): 927–31.

Brown, J.S., Collins, A. and Duguid, P. (1989) 'Situated cognition and the culture of learning', *Educational Researcher*, 18: 32–42.

Carrier, L.M., Cheever, N.A, Rosen, L.D., Benitez, S. and Chang, J. (2009) 'Multitasking across generations: multitasking choices and difficulty ratings in three generation of Americans', *Computers in Human Behavior*, 25: 483–9.

Crain, W. (2000) *Theories of Development: Concepts and Applications*, Englewood Cliffs, NJ: Prentice Hall.

Dewey, W. and Steinberg, H. (2001) 'Mood change through physical exercise in nine-to ten-year-old children', *Perceptual & Motor Skills*, 93(1).

Doğan, F. (2006) 'Video games and children: violence in video games', *Symposium Journal*, available at: http://www.yenisymposium.net/fulltext/2006%284%29/ys2006_44_4_1.pdf (accessed 29 October 2010).

Finkelhor, D., Mitchell, K.J. and Wolak, J. (2008) *First Youth Internet Safety Survey (YISS-1) {Dataset}*, available from National Data Archive on Child Abuse and Neglect website, at: http://www.ndacan.cornell.edu (accessed 3 November 2010).

Fox, A.B., Rosen, J. and Crawford, M. (2008) 'Distractions, distractions: does instant messaging affect college students' performance on a concurrent reading comprehension task?', *Cyberpsychology and Behavior*, 12: 51–3.

Frye, D., Zelazo, P.D. and Palfai, T. (1995) 'Theory of mind and rule based reasoning', *Cognitive Development*, 10: 483–527.

Genevieve, J. (2010) 'Young children's Internet use at home and school: patterns and profiles', *Journal of Early Childhood Research*, 8(3): 282–93.

Gross, E.F. (2004) 'Adolescent Internet use: what we expect, what teens report', *Journal of Applied Developmental Psychology*, 25: 633–49.

Healy, J. (1998) *Failure to Connect: How Computers Affect Our Children's Minds and What We Can Do About It*. New York: Simon & Schuster.

——(2004) *Your Child's Growing Mind: Brain Development and Learning from Birth to Adolescence*. New York: Broadway Books.

Heim, J., Brandtzaeg, P., Hertzberg, B. and Endstad, T. (2007) 'Children's usage of media technologies and psychosocial factors', *New Media & Society*, 9(3): 425–54.

Helmeke, C., Ovtscharoff, W., Poeggel, G. and Braun, K. (2001) 'Juvenile emotional experience alters synaptic inputs on pyramidal neurons in the anterior cingulate cortex', *Cerebral Cortex*, 11(8): 717–27.

Hickling, A. and Wellman, H. (2001) 'The emergence of children's causal explanations and theories: evidence from everyday conversation', *Developmental Psychology*, 57(5): 668– 81.

Juvonen, J. and Gross, E. (2008) 'Extending the school grounds? Bullying experiences in cyberspace', *Journal of School Health*, 78(9): 496–505.

Kalat, J. (1998) *Biological Psychology*, 6th edn, Pacific Grove, CA: Brooks and Cole Publishing Company.

King, A., Baumann, K., O'Sullivan, P., Wilcox, S. and Castro, C. (2002) 'Effects of moderate-intensity exercise on physiological, behavioral, and emotional responses to family care giving: a randomized controlled trial', *Journals of Gerontology: Series A: Biological Sciences & Medical Sciences*, 57A(1): M26–M37.

Leatherdale, S.T., Woodruff, S.J. and Manske, S.R. (2010) 'Energy expenditure while playing active and inactive video games', *American Journal of Health Behavior*, 34(1): 31–5.

Lee, S. and Chae, M.A. (2007) 'Children's Internet use in a family context: influence on family relationships and parental mediation', *Cyberpsychology & Behavior*, 10(5): 640–4.

Lenhart, A. and Madden, M. (2007) *Teens, Privacy & Online Social Networks*, Washington, DC: Pew Internet & American Life Project.

Lenhart, A., Purcell, K.S. and Zickuhr, K. (2010) 'Social media and young adults', a Pew Internet and American Life Project report, available at: http://pewInternet.org/Reports/2010/Social-Media-and-Young-Adults.aspx?r=1 (accessed 4 November 2010).

Livingstone, S. (2003) 'Children's use of the internet: reflections on the emerging research agenda', New Media & Society, 5(2): 147–66.

Lwin, M.O., Stanaland, A. and Miyazaki, A. (2008) 'Protecting children's privacy online: how parental mediation strategies affect website safeguard effectiveness', *Journal of Retailing*, 84: 205–17.

Lynda, M.A. and Kerry, J.D. (2010) 'Parents and adolescents making time choices: choosing a relationship', *Journal of Family Issues*, 31(11): 1419–41.

Medina, J. (2009) *Brain Rules*, Seattle, WA: Pear Press.

Mok, D., Carrasco, J.A. and Wellman, B. (2010) 'Does distance still matter in the age of the Internet?', *Urban Studies*, 47.

Newman, B. and Newman, P. (2001) 'Group identity and alienation: giving the web its due', *Journal of Youth and Adolescence*, 30(5): 515–39.

Nie, N. and Erbring, L. (2000) *Internet and Society: A Preliminary Report*, Stanford, CA: Stanford Institute for the Quantitative Study of Society.

Norris, T. (2010) 'Adolescent academic achievements, bullying behavior, and the frequency of Internet use', doctoral dissertation, available from Dissertations and Theses database. (UMI Number: 3405534).

Ophira, E., Nass, C., and Wagner, A.D. (2009) 'Cognitive control in media multitaskers', *Proceedings of the National Academy of Sciences of the United States of America*, 106(33), available at: http://www.pnas.org/content/106/37/15583.

Patchin, J. and Hinduja, S. (2006) 'Bullies move beyond the schoolyard: a preliminary look at cyberbullying', *Academy of Criminal Justice Sciences*, 4: 148–69.

Roberts, D.F., Foehr, U.G. and Rideout, V.G. (2005) 'Generation M: media in the lives of 8–18 year-olds', Menlo Park, CA: Henry J. Kaiser Family Foundation, available at: http://www.kff.org/entmedia/loader.cfm?url=/commonspot/security/getfile.cfm&PageID=51809 (accessed 15 November 2010).

Rodkin, P. and Hodges, E. (2003) 'Bullies and victims in the peer ecology: four questions for psychologists and school professionals', *School Psychology Review*, 32(3): 384–400.

Sapolsky, R. (1998) *Why Zebras Don't Get Ulcers*, New York: W.H. Freeman and Company.

Sefton-Green, J. and Buckingham, D. (1996) 'Digital visions: children's creative uses of multi-media technologies', *Convergence*, 2(2).

Subrahmanyam, K. and Lin, G. (2007) 'Adolescents on the Net: Internet use and well-being', *Adolescence*, 42(168): 659–72.

Sylwester, R. (1997) 'The neurobiology of self-esteem and aggression', *Educational Leadership*, 54(5): 75–9.

Tarpley, T. (2002) 'Children, the Internet and other new technologies', in D.G. Singer and J.L. Singer (eds) *Handbook of Children and the Media*, Thousand Oaks, CA: Sage, pp. 547–56.

Turkle, S., Taggart, W., Kidd, C.D. and Dasté, O. (2006) 'Relational artifacts with children and elders: the complexities of cybercompanionship', *Connection Science*, 18(4): 347–61.

Valkenburg, P.M. and Peter, J. (2007) 'Online communication and adolescents' well-being: testing the stimulation versus the displacement hypothesis', *Journal of Computer Mediated Communication*, 12(4), article 2, available at: from http://jcmc.indiana.edu.myaccess.library.utoronto.ca/vol12/issue4/valkenb (accessed 16 September 2011).

Veen, W. and Vrakking, B. (2006) *Homo Zappiens: Growing up in a Digital Age Network*, London: Continuum.

Wang, J., Iannotti, R.J. and Nansel, T.R. (2009) 'School bullying among adolescents in the United States: physical, verbal, relational, and cyber', *Journal of Adolescent Health*, 45(4): 368–75.

Wodarski, J.S. and Hedrick, M. (1987) 'Violent children: a practice paradigm', *Social Work in Education*, 10: 28–42.

Wolak, J., Mitchell, K. and Finkelhor, D. (2007) 'Unwanted and wanted exposure to online pornography in a national sample of youth Internet users', *Pediatrics*, 119: 247–57.

Ybarra, M.L., Mitchell, K.J., Wolak, J. and Finkelhor, D. (2006) 'Examining characteristics and associated distress related to internet harassment: findings from the Second Youth Internet Safety Survey', *Pediatrics*, 118(4): e1169–e1177.

Chapter 10

Online support in psychological and pedagogical practices

Bassam Aouil

Introduction

Modern information technology has been incorporated into almost all aspects of life. As a consequence of computer and internet accessibility, the percentage rise in their users, especially young people, has been observed.

My professional experience of psychological help for young people has suggested that young people prefer anonymous virtual counselling, psychological and spiritual support (i.e. 'virtual guardian angel'), where one can consult a psychologist any time about various issues, to direct contacts. However, this attitude is not a hindrance to real contacts (direct or the so-called live contacts), designed to offer help and support.

The study attempts to analyse some of the issues regarding online psychological help from the perspective of current challenges and expectations. It refers to methodology as well as practical and ethical issues and analyses the usefulness of this new type of help in enriching the achievements of clinical psychology and improving assistance tools.

Online psychological help

Online psychological help provides assistance via electronic media such as the internet, telephone, videotelephone, etc. Online help generally involves information and educational services as well as therapeutic services. The first type of services is provided by clinical psychologists to widely promote mental health by various indirect and direct methods. It can involve both contacting interested individuals (also parents, teachers and other professionals) and making use of websites to promote information on education and support. It mainly refers to opportunities and threats to the optimal development and functioning of people; it also concerns various social roles and situations (Aouil, 2004), whereas the therapeutic aspect of online psychological help refers to providing services for people in need via the internet and its tools such as e-mail, videotelephone, etc. (see also Caselnuovo *et al.*, 2001).

Therapeutic activity of online psychologists involves psychological and therapeutic services such as psychosocial support with elements of crisis intervention and psychological counselling, combined with psychoeducation and psychoprevention. Nonetheless, it is not actually traditional psychotherapy due to various possible errors in communication and specific interpersonal relationship between a therapist and client. In theory, however, the aim of a therapeutic effect of online psychological help is the same as the aim of standard psychotherapy, which involves eliminating and alleviating consequences of functioning disorders as well as developing a patient's personality and enhancing their mental strength (Aouil, 2008).

The terminology and principles of online psychological help are directly connected with the main tasks of clinical as well as developmental and educational psychology. The pioneer of online help was Ivan Goldberg (see Young, 2005). In 1993, he presented information on depression treatment on his *Depression Central* website and initiated a support group for people suffering from a disease termed 'Walking In Darkness'. Since then, numerous other forms of internet-based help have been introduced, including support groups such as discussion lists and forums (McKenna and Bargh, 1998; Wallace, 2001; Chang and Yeh, 2003; Winefield, 2005; Ybarra and Eaton, 2005) as well as attempts at online psychotherapy regarding specific disorders (Rochlen *et al.*, 2004).

In Poland, there has been a considerable increase in websites and bulletins devoted to online psychological services. In comparison to years 2002–3 when I, as one of the first in Poland, launched an online psychological help service at the Academic Centre of Psychological Help (www.acpp.ab.edu.pl) at the then Kazimierz Wielki Bydgoszcz Academy (psych.ab.edu.pl/cpp, at present www.acpp.ukw.edu.pl). I have also offered a similar service on the portals (www.zdrowemiasto.pl and www.ptt-terapia.pl), while my most recent work is a portal for grammar and secondary school students (www.psycho-help.pl) (since 2003, over 7000 individuals have been helped by me and my team through online psychological counselling), an increase in companies and individuals providing internet counselling has been observed (the Google query 'online psychological help' will currently return over 330,000 results on 29 May 2010). Although European and Polish statistics have been less impressive than American ones, the numbers have definitely been increasing. In one year alone, 1998, 22 million people were provided with health information via websites (Rabasca, 2000). An estimated 5 per cent of websites discuss mental health (Eysenbach and Kohler, 2004).

Online therapeutic services in clinical practice: this author's perspective

As far as psychological help provided via e-mail or other messengers (IRC, chat, etc.) is concerned, online therapeutic contact can be established by internet telephone conversation (voice-over or video conference), and it is essential that a client consults a therapist when they need it, as frequently and as long as they decide to do so. It is clients who initiate contact, which may concern developmental, personality, social, professional, emotional-sexual and family relationship problems. A psychologist-therapist may suggest the frequency or number of consultations but it is always clients who decide.

Considering my own practice, it is my belief that there are actually no limits to what problem can be discussed via the internet (despite the absence of physical contact) and whether this form of contact will be effective in a given case, because actually no psychologist can be absolutely sure about their decision on a specific approach or work method. One can never be certain that the client will continue to attend therapeutic meetings. Therefore, it is not the nature of problems that imposes the limit but a patient's needs, communication capabilities and the ability to describe one's difficulties. When it comes to online therapeutic contact, motivation and confidence in this form of help become even more important than in the case of classical psychotherapy as it is not an easy task for online therapists to strengthen the motivation of distance patients. Online therapists can use the same approaches and techniques as traditional therapists, that is cognitive, behavioural and psychoanalytical methods.

However, where immediate physical or medical intervention for individuals with major problems is necessary, e-help can only be used as a temporary solution and cannot replace traditional psychotherapy.

In my experience, it is possible to provide online help for behaviour disorders and adjustment difficulties regarding, for instance:

1 coping with difficult or crisis situations concerning personal, family relationships, professional and social life issues;
2 coping with one's own or loved ones' illness and its consequences;
3 solving problems concerning addiction to alcohol or drugs;
4 solving relationship and sexual problems;
5 solving personal problems concerning one's system of values;
6 solving educational and school problems;
7 preventing dangers and solving current problems in order to help develop one's self-reliance, responsibility and ability to make decisions;
8 preventing dangers and providing help concerning anxiety behaviour, aggression, self-aggression, etc.;
9 shaping adequate self-appraisal, learning self-acceptance and ways to be yourself;
10 choosing and fulfilling life aims;
11 shaping sexual or personal identity.

Moreover, therapeutic services involve the support as well as psychosocial and professional rehabilitation of ill and disabled people by various means of *mutual help and online psychoemotional support*, which are also designed for people with addiction and violence problems.

My clinical experience concerning both young people and adults has confirmed that online counselling and help can be an additional or alternative form of assistance, compared to traditional help. It is an ideal method of establishing contact for clients who wish to stay anonymous. Not only can the internet be used for contacting clients but also for improving cooperation between psychologists in order to, for instance, seek expert secondary opinions. Considering the wide availability of the internet, online psychological help makes it possible for therapists and patients to maintain contact even after the completion of therapy. Most importantly, it is an indispensable prevention and information tool.

Popularity of online psychological services

According to my own survey on online help, conducted with a group of 750 young people (school and academic students) in years 2003–9, more and more school and academic students recognize the existence of online psychological services and it is they who most frequently decide to make use of it (in 2005 – 11 per cent, in 2006 – 19 per cent, in 2008 – 28 per cent, in 2009 – 44 per cent) when personal, family, social, relationship or sexual difficulties arise (Aouil, 2008).

Taking all results into consideration (see Aouil, 2004, 2008, 2009), it can be concluded that there is a need for this type of help and the interest in online psychological help has gradually been increasing. Clients most frequently value anonymity, comfort and availability of help whenever they experience difficulties.

Contemporary psychologists should make an effort to adjust conditions, stages, methods and forms of work and contact to patients' individual needs. Numerous people consider online psychological help to be a trustworthy form of help and they recognize its effectiveness (see Aouil, 2004, 2008, 2009). Such processes create opportunities for the improvement of healthcare and better feedback as well as for overcoming resistance to consult therapists.

According to numerous authors, such as Grohol (2004), one of the major reasons behind using online psychological help is mainly the opportunity to receive advice and support whenever needed. It particularly applies to inhabitants of distant or rural areas who may not have access to well-qualified specialists or mental health centres. Supplied with the opportunity to use online help services, they are given the chance of contacting professionals and online support groups.

Other people may have difficulty with social relations in real life. Such people do not need to feel lonely – they can compare feelings and thoughts with others and consequently receive support and life help. In the process of seeking communication opportunities and professionals to consult, one should choose online help. Finally, for numerous people there is still a lot of stigma attached to mental illnesses. Online services provide opportunities for removing much of the stigma from these problems (Grohol, 2004).

Qualifications of an e-therapist versus the effectiveness of help

Considering some of the above-mentioned theoretical and clinical aspects of online psychological help, such clinical practice appears to be undoubtedly important and advisable and the number of people willing to use online help is growing daily.

Apart from the earlier mentioned arguments, the reasons behind the necessity to discuss the qualifications of an e-psychologist are, on one hand, survey results concerning the opinions of various groups of people, including psychologists, about the significance of online psychological help in clinical practice and, on the other hand, the alarming attitudes of some professionals who outwardly provide online psychological help or e-therapy and actually 'experiment' with clients, trying to help without the use of professional background. Although I do not have assessment tools regarding one's theoretical or professional qualifications for working directly with clients, I would like to remark that the efficiency and effectiveness of online psychological help do not often depend only on the contents of help but also on the manner of providing it.

The differences between classical and online help are mainly the following:

1. characteristic features of distance communication;
2. limitations resulting from the impossibility of observation and the absence of other immediate and auxiliary interaction tools, important for therapists;
3. wide range of ages of clients (my youngest e-client was 9 and the oldest was 72 years old);
4. wide range of problems raised by e-clients.

Similar to direct contact, for therapeutic change to occur, there are necessary conditions, described by C. Rogers (1991), that also have to be satisfied in regard to online help. Based on my own clinical and research experience, let me analyse the structure of qualifications of an e-therapist in order to determine development areas.

Online psychological help has essentially been designed to do the following:

1. help provide revealing insight into oneself and one's situation;
2. help make adequate assessment of one's individual abilities, environment, source of stress and helplessness;
3. gain a better understanding of oneself and one's own behavioural motives (conscious and unconscious) and defence mechanisms;

4. arrive at one's own conclusions and develop the knowledge of oneself and one's environment;
5. make it easier for clients to take actions to achieve better adjustment and solve problems on the basis of their hierarchy of values and beliefs.

In addition, it has to be recognized that there are limits concerning online psychological help. They mainly refer to certain ethical standards. This seems to be the weakest point of e-help practice, both in countries which have had e-help tradition and in Poland. The ethical side also definitely concerns the e-therapist certification system and professional responsibility towards e-clients.

Considering the recommendations of American organizations which were the first to provide online psychological help (American Counseling Association, 1999; National Board for Certified Counselors, 2001), the observations of researchers in the field (e.g. Ragusea and VandeCreek, 2003; Rosińska-Włodarczyk, 2007) and my own experience, the following competence standards regarding effective online help can be established:

1. considerable knowledge of psychological theories and professional qualifications;
2. considerable knowledge of theoretical and practical principles of distance psychological help;
3. considerable knowledge of communication technologies and ability to apply them;
4. knowledge of online communication methods and techniques as well as basic principles behind internet linguistic and non-textual communication;
5. considerable knowledge of the code of ethics suggested by foreign psychological societies which have incorporated online help into their activity;
6. experience of using the media in psychological practice and experience of information management by the use of media as well as the ability to use electronic information resources in the process of communication with clients;
7. ability to identify and diagnose a client's problems and specify the current needs of clients from various socio-cultural environments;
8. knowledge of institutions providing psychological help, including online institutions, and their competence;
9. knowledge of national and European network of consultants;
10. ability to establish verbal and non-verbal contact with a client as well as maintain the balance between a monologue and dialogue by means of non-verbal communication;
11. willingness to attend training and supervision sessions. Supervision is a psychotherapeutic method, rather obligatory for psychotherapists. It involves regular meetings between a therapist and another specialist certified as a supervisor. During a meeting, a supervisor and psychotherapist discuss

> selected cases regarding clients and other issues connected with the duties of a psychotherapist. Therefore, these meetings aim to provide assistance for psychotherapists and enhance the quality of therapeutic services.

It should be noted that not all psychologists who have graduated from psychological studies and have even obtained a licence for a given therapeutic competence are automatically prepared for this type of work. Therefore, considering the need to systematize a newly emerging market in order to provide high-quality online psychological services, at this stage in the development of online therapeutic services, training, supervision and associations of online psychologists-consultants have necessarily to be discussed.

Moreover, it has to be recognized that certain remarks about the practice of online psychological help result from a specific code of ethical conduct. At present, this aspect is one of the weakest points of the practice of e-help, both in countries which have a tradition of psychological help and in Poland. The ethical side definitely concerns both the e-therapist certification system and professional responsibility towards e-clients.

Stages and characteristics of online psychological help

The first stage involves establishing online contact due to the initiative of a person who needs help. It usually begins with an *application* made on the portal or submitted to a psychologist. The request is mostly made via e-mail. On a well-structured online help portal or website there will be an application form requesting basic information about a client (and ensuring the anonymity of clients in terms of personal details) and giving the opportunity to describe one's problem. This is usually a detailed description of ailments and difficulties experienced by a given client. Some clients focus mainly on a problem itself, ignoring formal details or providing false information for fear of being identified. An example is shown below:

> My name's Monika, I come from Wrocław and my problem's domestic violence. I don't want to talk to the school psychologist because I know I'd be recognized and my girl friends would find out about these visits. I've tried to cope myself but it's come to nothing. I've decided to write to you because I've had enough of this life. Please help.

More detailed messages probably result from the need for immediate help or concern a crisis situation. An e-therapist initiates the preliminary stage of establishing therapeutic contact and begins psychoeducation regarding problems that have been raised; a therapist also asks additional questions to gain a better understanding of a problem and a client's personal and social situation. At this stage it is also necessary to determine what kind of help can be provided and which tool to use.

After an e-client has answered a therapist's questions and the type and range of help have been determined, the next stage – therapeutic contact – begins. A contract is an agreement between an e-therapist and an e-client, aimed at determining the conditions of help, enhancing the security of both parties and preventing potential misunderstandings about the process of e-therapy or its results and effects.

In one of the initial e-mails, an e-therapist presents the terms of the contract and confirms them with the e-client. Both have the right to negotiate the terms of the contract before the final approval.

The contents of the contract can be negotiated until the approval, especially in regards to the frequency of contact, waiting time for an e-mail reply, length of conversation via Skype or telephone or the possibility of sending SMS messages or talking via mobile phone. If it is supposed to be paid help, such conditions have to be precisely determined.

Essential elements of the help process include *identification of a problem* and the factors behind the problem and finally the *search for the solution to the problem*. There are essentially four stages of online psychological counselling:

1 *Problem identification*: Identification of a difficulty experienced by an e-client – usually described as difficulty in fulfilling specific tasks concerning life or development or a new situation.
2 *Identification of factors contributing to the persistence of a problem and to difficulties encountered during the attempts to solve it*: In terms of feelings experienced by clients (e.g. fear), impeding the application of one's skills, causing the absence of certain skills or inability to use them.
3 *Search for alternative solutions to difficulties experienced by e-clients*: The solutions have to be specific, applicable to a client's circumstances and have to make it possible to experience an immediate change (even the slightest improvement is excellent).
4 *Completion* : Contact is made again after short or long periods of time to support and enhance the client's efforts to make changes; it is also an opportunity to modify the earlier accepted strategy.

Therapy effects usually arise from a supporting atmosphere (trust, acceptance and support) and therapy mechanisms (when clients talk about problems and describe them). Therapeutic mechanisms also result in: filtering and reduction of negative emotions and experiences as well as elimination and alleviation of symptoms.

The aims of e-help govern the direction of actions and the choice of help strategies adopted by an e-therapist. Similarly to psychological counselling, the strategies can be directive (instructional strategy and confrontational strategy) or non-directive (catalyzing strategy, discharge strategy, information strategy, support strategy).

Essential duties of an e-therapist involve providing help in the identification and specification of an objective, or objectives, and construction of their hierarchy. An e-therapist teaches a client how to participate in e-therapy, and adapts to the type of relation established by a client. For direct contact with clients, each e-therapist uses their own therapeutic skills and abilities. Their skills become resources which enrich the process of e-help.

The future of online psychological help

There are numerous disputes among psychologists and psychotherapists over e-therapy, that is, psychological service provided via the internet. Negative attitudes often result from ignorance about this category of support. A number of specialists cannot imagine internet therapy, pointing to differences between online help and direct therapy sessions, held in specialist offices. The results emphasize the necessity to spread the knowledge of online psychological activity, especially concerning rules of application.

A serious obstacle in the path of the development of e-help is the absence of professional and ethical principles. The commonest accusations made against online help are the following:

1. It is 'unethical' and 'unprofessional' due to the lack of legal and professional rules; their absence can pose threats to clients.
2. Communication based 'on text' or 'without the possibility of observation and direct contact with a client' cannot be regarded as psychotherapy or even counselling owing to various limitations resulting from distance communication.

What measures can be taken then? Without doubt nobody will ever halt the development of online psychological services, especially because of their growing popularity and clients' expectations.

Psychological e-help can definitely be considered professional, but it has to be emphasized that it is not a good solution for everybody. Moreover, it cannot fully replace direct help. It is on the professional qualifications of psychologists/online consultants and online psychologists' code of ethics that the range of services, including psychoeducation, prevention, support, intervention, counselling or therapy, will depend. After all, refusing to help people in need seems unethical.

Certification, training, supervision

In order to establish professional standards of psychological e-help, it is necessary to promote training and supervision for e-therapists so that they can be awarded professional certificates. I have been the head of the Section of Counselling and Media Presentations of the Polish Psychological Society which has made a comprehensive suggestion, according to which, among other things, candidates applying for the Certificate of Online Consultant have to collect 300 obligatory points (out of 500) and subsequently pass an exam.

Association of consultants

The creation of professional standards of online psychological services not only involves investment in team qualifications and technology but also legal protection and organizational background. An association of e-therapists can provide such protection and basis for further research on the improvement of work methods, with the use of clear, lawful and ethical basics. My suggestion is to establish the 'Online Psychologists' Association' or the 'Polish Online Psychologists' Association' (Polskie Stowarzyszenie Psychologów Online – PSPO).

From the perspective of the world development of the market of online services, it is not a completely original idea. Throughout the very short history of online help there have been such organizations before. The first one was the ACA (the American Counseling Association). Other organizations of this type tried to issue a consultant licence. One of the first ideas was a consultant diploma issued by the National Board of Certified Counselors (NBCC). Another example would be the most popular organization for e-therapists known as ISMHO, which attempts to point to the major importance of online psychological help for academic research.

Ethical context

Establishing the code of ethics is an indispensable basis for the development of online psychological services. The plan of the code of ethics for e-therapists is under preparation and it includes the following assumptions:

1 possibilities and limitations concerning the positive effects of online help;
2 situations providing opportunities for therapeutic contact between an e-therapist and client;
3 limits of online diagnosis;
4 client's right to anonymity and comfortable communication tools as well as a therapist's obligation to perform overt actions;
5 situations when direct contact with a psychologist or psychiatrist is necessary;
6 decisions concerning payment for services.

The above discussed concept corresponds to the principles of Health On the Net Foundation (Health on the Net Foundation, 2001) and the suggestion made by Ron Kraus (2004), who points to the necessity of establishing such a code of ethics, although he is completely aware of the fact that the code will not solve all dilemmas and conflicts but still it will be the basis and guarantee that a psychologist will attempt to provide services of the highest possible quality.

References

Aouil, B. (2004) *Psychologia pomocy Online*. Bydgoszcz: Magraf s.c.

——(2008) *Pomoc psychologiczna Online: Teortyczne podstawy i praktyczne wskazówki*, Bydgoszcz: MW.

——(2009) *Pomoc Psychologiczna Online – teoretyczne podstawy i praktyczne wskazówki*, 2nd edn, Bydgoszcz: MW.

Bell, V. (2007) 'Online information, extreme communities and internet therapy: is the internet good for our mental health?' *Journal of Mental Health*, 16(4), 445–57.

Caselnuovo, G., Gaggiolo, A. and Riva, G. (2001) 'CyberPsychology meets clinical psychology: the emergence of e-therapy in mental health care', in R. Giuseppe and C. Galimberti (eds) *Towards CyberPsychology: Mind, Cognitions and Society in the Internet Age*, Amsterdam: IOS Press, pp. 229–52.

Chang, T. and Yeh, C.J. (2003) 'Using on-line groups to provide support to Asian American Men: racial, cultural, gender, and treatment issues', *Professional Psychology: Research and Practice*, 34: 634–43.

Eysenbach, G. and Kohler, C. (2004) 'Health-related searches on the Internet', *Journal of the American Medical Association*, 291(24): 29–46.

Grohol, J.M. (2004) *The Insider's Guide to Mental Health Resources Online*, New York: The Guilford Press.

Health On the Net Foundation (2001) *HONcode*, available at: http://www.hon.ch/HONcode/Guidelines/guidelines_pl.html (accessed 5 May 2007).

Kraus, R. (2004) 'Ethical and legal considerations for providers of mental health services online', in R. Kraus, J.S. Zack and G. Stricker (eds) *Online Counseling: A Handbook for Mental Health Professionals*. Boston: Academic Press, pp. 123–44.

Kraus, R., Zack, J. and Stricker, G. (2004) *Online Counsellling: A Handbook for Mental Health Professionals*. Waltham, Mass.: Academic Press.

McKenna, K.Y.A. and Bargh, J. (1998) 'Coming out in the age of the Internet: identity "demarginalization" through virtual group participation', *Journal of Personality and Social Psychology*, 75: 681–94.

National Board for Certified Counselors (2001) 'The practice of Internet counseling', available at: http://www.nbcc.org/ethics/webethics.htm (accessed 4 April 2002).

Rabasca, L. (2000) 'Self-help sites: a blessing or a bane?', available at: http://www.apa.org/monitor/apr/00/selfhelp.html (accessed 12 June 2008).

Ragusea, A.S. and VandeCreek, L. (2003) 'Suggestion for the ethical practise of online psychotherapy', *Psychotherapy: Theory, Research, Practice, Training*, 40(1–2): 94–102.

Rogers, C.R. (1991) Terapia nastawiona na klienta. Grupy spotkaniowe, Wrocław: Thesaurus Press.

Rosińska-Włodarczyk, P. (2007) Standardy etyczne w pomocy online, in B. Aouil (ed.) *Internet in Psychology: Psychology of the Internet,* Torun: A. Marszalek Press.

Wallace, P. (2001) *Psychologia Internetu*, Poznań: Rebis.

Williams, S. (2000) *Facts and Figures*, available at: http://www.apa.org/monitor/apr00/facts.html (accessed 20 May 2009).

Winefield, H.R. (2005) 'Support provision and emotional work in an Internet support group for cancer patients', *Patient Education and Counselling*, 62: 193–7.

Ybarra, M.L. and Eaton, W. (2005) 'Internet-based mental health interventions', *Mental Health Services Research*, 7: 75–87.

Young, K.S. (2005) *Caught in the Net*, New York: Wiley and Sons.

Chapter 11

Online counselling for enhancing relationships

Kevin Glasheen and Marilyn A. Campbell

Most young people are students attending school where their peer relationships impact on their educational outcomes. At the end of the first decade of the twenty-first century, young people also depend on the communication technologies to assist them with their developing relationships. Increasingly this connectedness to the peer group is being carried out both in the physical and virtual worlds as evidenced by the rapid increase in the membership of social network internet sites. However, schools seem reluctant to implement educational processes that utilize the social networking sites that young people inhabit. In fact, schools seem to be only emphasizing the dangers of such social networking sites and the positive benefits of technology for enhancing peer relationships do not seem to be acknowledged. Online school counselling could be one way for schools to promote the positive uses of technology to contribute to young people's education and to enhance their social relationships and connectedness.

Importance of technology in young people's lives

The Internet and communication technologies play an ever increasing role in the social lives of young people in Western societies at present. Young people treat the mobile phone as an essential necessity of life and often prefer to use synchronous chat such as *Windows Live Messenger* (MSN) to communicate with their friends (Campbell, 2005) as well as accessing social networking websites such as Facebook and MySpace. The number of young people with Facebook accounts has increased dramatically in the past few years.

It has been 20 years now since the internet was first introduced into Australia, and since then over one half (12 million) of Australians have become connected. Technologies such as personal music devices, personal organizers, Iphones, Ipads, portable computers, and remote internet connection have created a seamless on and offline social life for most adolescents. In a recent Australian study, 626 students ranging in age from 10–16 years old reported they spent an average time of 6 hours a week socializing online

(Bonetti *et al.*, 2010) which accords with the 49 minutes a day found in 2007 (Australian Communications and Media Authority, 2007).

Importance of relationships in young people's lives

The importance that young people attribute to their peer relationships is well known. In the past hundred years adolescence has become a time of separating from the family and creating a youth culture which distinguishes each generation from the previous one. Thus youth culture means many adults are now excluded and adolescents spend most of their time with peers (Steinberg, 2005).

As part of identity development, membership of the peer group provides the young person with the opportunity to explore various types of relationships and learn how to regulate the mechanisms of emotions associated with these relationships. The skills young people learn in dealing with their emotions at this stage have been found to have an impact on the quality of their later adult relationships (Dhariwal *et al.*, 2009). The role of romantic relationships in the adolescent years is sometimes regarded with little importance by adults, however, recognition of how such relationships are significant for a young person's sense of belonging to the peer group and their development as healthy adults is gradually being acknowledged (Collins, 2009).

As the positive emotional health of the young person is largely dependent on the presence of protective factors such as positive self-esteem and connectedness to family and school (McAnarney, 2008), it could be argued that connectedness to a social network such as the peer group is also crucial for the young person's mental health. This connectedness is reflected in the quality of peer relationships.

Importance of schools in building relationships

As most young people are students within a school system, it is appropriate that as educational facilities, schools assist individuals in the development of their social skills. School is the place where many form lifelong friendships and find even lifelong partners. One Italian study indicated the importance of adolescent relationships as a means of developing skills in dealing with their emotions, especially in romantic relationships, and how these have an impact on the quality of later adult relationships (Dhariwal *et al.*, 2009).

How can schools better develop peer-to-peer relationships through online technology?

However, online communication with peers is perceived by teachers as being time-wasting. Perhaps this is because the adults in schools assume that there are parallel universes of the 'real' world in the physical world and the 'virtual' or unreal world in cyberspace. This is shown in many ways but one of the most visible ways is the banning of most popular online networking sites in schools such as Facebook. One of the reasons given is that there are too many 'dangers' associated with networking sites. Or alternatively adults see the physical world as more important. A focus group of school counsellors, when asked about online communication with students, assumed that the physical presence of a student was more important than online communication (Glasheen *et al.*, 2010). However, the physical and the virtual world are psychologically connected; they are not actually separate universes. In the young person's reality, transition between the two is seamless.

This distrust and consequent banning of peer-to-peer use of technology do not fit well with the school's aims of engendering a friendly school community with students helping other students. The school does not consider friends sitting together or playing together at break times as time-wasting but this seems to be the case only in the 'real' world and not online. Being friendly online is somehow considered as second rate and only 'true' friends can be made in the physical world. School systems are therefore reluctant to implement educational processes that utilize the social networking sites that young people inhabit. This could be due to educational authorities being suspicious of the educational benefit of using such sites and being cautious about the implications of students being exposed to certain internet sites and the potential of giving online predators access to students. This has been justified on the basis of 'duty of care' (Notley, 2008) but the positive benefits of technology for enhancing peer relationships do not seem to be acknowledged. Schools are therefore placed in a difficult position in trying to control the negative and destructive uses of electronic media but also to find ways to promote the positive uses of technology to contribute to young people's education and to enhance their social relationships and connectedness.

How can positive use of technology promote relationships?

One of the positive benefits of new technologies is that young people have a means of networking with others. This can encourage help-seeking behaviours especially in relationship problems such as bullying and cyberbullying which adversely affect mental health. The new form of bullying, cyberbullying, actually uses the technology in a detrimental way. Cyberbullying has a variety of attributes that appear to accentuate the impact of the bullying behaviour, potentially including a wider audience, anonymity, the more enduring nature of the written word and the ability to reach the target at any time and in any place including previously considered safe havens such as the target's home. The comments delivered online can also be much nastier than would normally be said in a face-to-face situation. Preliminary research suggests that compared to face-to-face bullying, cyberbullying impacts on students' mental health more than face-to-face bullying. A recent study by Campbell *et al.* (2010) found that cyberbullying victims reported more anxiety, depression, and behavioural difficulties than victims who were bullied in person. These severe consequences urgently require evidence-based intervention research.

Young people are often reluctant to seek professional help for personal and psychological problems (Rickwood *et al.*, 2005). The barriers to help-seeking include stigma, fears regarding confidentiality, embarrassment and lack of appropriate youth-friendly services as well as fear of retaliation from bullies (ibid.). This is of concern, especially for young men who experience poorer educational outcomes, and higher rates of incarceration, illness and completed suicides than young women in Australia (Sawyer *et al.*, 2001).

Why online counselling?

Online counselling refers to the synchronous use of computer-mediated communication in the counselling process. In practice, a counsellor conducts a counselling interaction with the client within a live, real-time chat room. Online counselling could become a critical and age-appropriate way of extending the provision of school counsellor support to adolescents who have increasingly embraced technology as a means of communication and support.

Online communication is favoured by young people because it is often not as threatening as face-to-face communication, especially when dealing with sensitive or emotionally charged issues. When communicating online, participants can be less inhibited. This has been coined as the *online disinhibition effect* which can be either toxic or benign (Suler, 2005). The former contributes to the veracity of bullying online and the latter assists the person to divulge personal information and details they would otherwise not reveal. Traditionally the analyst sat behind the client who reclined on the couch thereby avoiding the need for the client to engage physically with the therapist. Even if the person knows the person to whom they are messaging, this physical invisibility assists those who are more vulnerable or less socially confident. As the identity of the individual is partially hidden, it is conducive to a therapeutic relationship and often more comfortable for the young person. The use of text is also not a new phenomenon when communicating at an intimate level. In times past, letters were often a means of expressing emotions and feelings to distant friends or loved ones. When the written word can be used synchronously, it combines the benefits of immediacy and the online disinhibition effect which has potential to build or repair relationships. These online benefits could be used in online counselling when there is a crisis in a young person's relationships, resulting in bullying or harassment.

For adults, online counselling is widely available, which has seen the growth of publications and texts for practitioners (Evans, 2009; Anthony and

Nagel, 2010). Online counselling for adults has shown a rapid increase in Australia (Gedge, 2002) however, there are not as many online counselling services for young people.

Some community-based services provide online counselling which are open to young people, though these are often restricted to specific issues such as drug and alcohol use. As discussed earlier, online counselling is less intrusive for young people who are often reluctant to seek help as they feel they might lose control of their emotions. Boys, who often fall into this category, seem to appreciate the less intrusive nature afforded online. Additionally, they are afraid of being seen as weak and therefore are concerned about being teased (Glasheen, 1998). Using computers may be particularly empowering for people who feel intimidated initially and adolescents who are familiar with technology and are averse to face-to-face counselling could benefit by online counselling (Collie *et al.*, 2002).

Additionally, there is also less stigmatism associated with online counselling, as no one knows if the student is accessing help. This is in direct contrast to most school counselling situations where students must get passes to be out of class to access the counsellor. This is part of the client having more control and the levelling of power differences between counsellor and client on line (Murphy and Mitchell, 1998). Online counselling also has the advantage of more flexibility. Instead of students missing classes, they can access the school counsellor at other times. Online counselling also means that there is a permanent record of the sessions. The records from e-mail exchanges can be beneficial for a client to chart their progress, for counsellor training, supervision and research (Oravec, 2000).

Trial of online counselling in a school

The difficulties faced by young people, the fact that many are reluctant to seek help, together with the popularity of technology use in this population led to a trial of a limited service of online counselling in one Australian secondary school (Glasheen and Campbell, 2008).

As the online service was used, it became obvious that in most cases, students followed up the online appointment with subsequent 'face-to-face' consultations with the counsellor. The trial found that a diverse group of students used the online mode. Students with disabilities appeared to be less inhibited online and possibly they were less self-conscious of their impairments compared to face-to-face discussions with the counsellor. Boys were the main users and some students found a combination of 'face-to-face' and online modes beneficial depending on the presenting issue at the time. Typical cases included the following:

- *It's too personal to talk to someone about:* 'Ben' requested an online appointment on the last day of term. During the first online interaction with the school counsellor, 'Ben' asked questions about sexuality and disclosed that he was same sex attracted. This allowed him to assess if it was 'safe' to discuss such issues with the counsellor. He continued with a number of online appointments before he agreed to further discuss this and other issues in face-to-face counselling subsequently arranged through the online process.

- *I can be like everyone else*: 'Darren' was confined to a wheelchair. His physical impairment was severe. He was not forward in seeking counselling assistance and generally he limited his verbal communication with the school counsellor. However, once he was introduced to the online facility, 'Darren' made a number of online appointments to discuss his life goals and his hopes for the future.
- *Can we discuss this together?* Brian and Chad were having difficulties with a particular teacher. They didn't want to talk about it but wanted to use the comic chat to help them communicate. They asked for an online meeting together. They used the library computers for a lunchtime meeting with the counsellor. They were able to discuss openly what was happening and share ideas with the counsellor on how best to work through the conflict.
- *I'm not supposed to have problems*: 'Kyle' was in a leadership role in the school. As such, he was reticent to seek out the school counsellor. However, when his home situation started to impinge on his ability to carry out the duties expected of him, he contacted the school counsellor initially via e-mail. Through this process, a synchronous meeting was arranged and a number of issues were discussed outside school hours online. Kyle chose to follow up some of these issues with further face-to-face counselling.

Conclusion

Adolescence is the time in the young person's development when the importance of the peer group is paramount. The relationships within the peer group are important as a preparation for later significant relationships and partnerships in life. At the end of the first decade of the twenty-first century young people are dependent on the communication technologies for the instigation, nurturing and maintenance of these relationships. Though online communication is integral to the individual being connected to their peer group, the technology can be used to fracture relationships through cyberbullying and abuse.

Schools have long been considered important agents of socialisation, and as such have taken on the role of social skilling students. However, there has been a hesitancy to use online technology and social network sites to foster socialization of young people. Though some initiatives have been taken to train cybermentors, there has been a lack of peer skilling programs which aim to promote students supporting students online. The challenge of using online technology to foster peer relationships now confronts school administrators and educators.

Until now, many students who experience conflict and difficulties in their peer relationships have sought face-to-face assistance from school counsellors. The innovation of school counsellors providing online counselling to students has been piloted with promising results. Early indications suggest that fractured relationships can be supported by the medium through which many were initially formed and later maintained. Especially for those students who are reticent to seek traditional forms of help, communication technologies can provide online support. The opportunity to resolve conflict and repair relationships is now possible through the availability of online counselling in schools and highlights the potential to enhance these relationships through using this technology in a positive manner.

References

Anthony, K.N. and Nagel, D.M. (2010) *Therapy Online: A Practical Guide*, London: Sage Publications Ltd.

Australian Communications and Media Authority (2007) *Media and Communications in Australian Families*, Canberra: Author.

Bonetti, L., Campbell, M.A. and Gilmore, L. (2010) 'The relationship of loneliness and social anxiety with children's and adolescents' online communication', *CyberPsychology and Behavior and Social Networking*, 13: 279–85.

Campbell, M.A. (2005) 'The impact of the mobile phone on young people's social life', paper presented at the Social Change in the 21st Century Conference. Brisbane, Australia, October.

Campbell, M.A., Spears, B., Slee., P., Butler, D. and Kift, S. (2010) 'Consequences of cyberbullying', paper in preparation.

Collie, K., Cubranic, D. and Long, B.C. (2002) 'Audiographic communication for distance counselling: a feasibility study', *British Journal of Guidance & Counselling*, 30: 269–84.

Collins, W.A. (2009) 'Adolescent romantic relationships', *Annual Review of Psychology(0066–4308)*, 60: 631–52.

Dhariwal, A., Connolly, J., Paciello, M. and Caprara, G.V. (2009) 'Adolescent peer relationships and emerging adult romantic styles: a longitudinal study of youth in an Italian community', *Journal of Adolescent Research*, 24(5): 579–600.

Evans, J. (2009) *Online Counselling and Guidance Skills: A Practical Resource for Trainees and Practitioners*, London: Sage Publications Ltd.

Gedge, R. (2002) 'Online counselling services in Australia: the challenges of a new vehicle for an old art', available at: www.ausweb.scu.edu.au/aw02/papers/refereed/gedge2/paper.html (accessed 7 July 2007).

Glasheen, K.J. (1998) 'Adolescents' perceptions of counselling and the implications for the development of masculine identity of adolescent males', unpublished thesis, QUT, Brisbane.

Glasheen, K. and Campbell, M.A. (2008) 'School counselling launches into cyberspace: an action research study of a school-based online counselling service', paper presented at the AARE International Education Research Conference, Brisbane, Australia, December.

Glasheen, K., Campbell, M.A. and Shochet, I. (2010) 'Hacking into the minds of school counsellors', paper in preparation.

McAnarney, E.R. (2008) 'Adolescent brain development: forging new links?', *Journal of Adolescent Health (1054–139X)*, 42(4): 321–3.

Murphy, L. and Mitchell, D. (1998) 'E-mail as therapy', *British Journal of Guidance and Counselling*, 26: 21–33.

Notley, T.M. (2008) 'Online network use in schools: social and educational opportunities', *Youth Studies Australia*, 27(3): 20–9.

Oravec (2000) 'Online counselling and the Internet: Perspectives for mental health care supervision and education', *Journal of Mental Health*, 9, 121–36.

Rickwood, D., Deane, F.P., Wilson, C.J. and Ciarrochi, J. (2005) 'Young people's help-seeking for mental health problems', *Australian e-Journal for the Advancement of Mental Health*, 4(3), Supplement.

Sawyer, M.G., Arney, F.M., Baghurst, P.A., Clark, J.J., Graetz, B.W., Kosky, R.J., *et al.* (2001) 'The mental health of young people in Australia: key findings from the child and adolescent component of the national survey of mental health', *Australian & New Zealand Journal of Psychiatry*, 35: 1002–14.

Steinberg, L. (2005) *Adolescence*, New York: McGraw-Hill.

Suler, J. (2005) 'The online disinhibition effect', *International Journal of Applied Psychoanalytic Studies*, 2(2): 184–8.

Part III

Prevention of cyberbullying

Chapter 12

Peer education intervention

Face-to-face versus online

Ersilia Menesini and Annalaura Nocentini

Introduction

In recent years, the use of new communication technologies has become a major component of adolescents' social lives in Italy (Menesini *et al.*, in press). Although young people frequently use technologies, they may not necessarily understand all the risks involved and the strategies to keep their experience of technology safe. Positive use of ICT can be a way to improve young people's skills to cope with the problem: there is empirical evidence that the positive use of ICT can be beneficial to child cognitive and social development and can enhance his/her capacity to deal with the risks of adverse situations and contexts.

Cyberbullying can be one of these online risks and it is becoming an issue of great concern amongst psychologists, educators and adults working with the young. Following the definition of traditional bullying, it has been defined as: 'an aggressive, intentional act carried out by a group or individual, using electronic forms of contact, repeatedly and over time, against a victim who cannot easily defend him or herself' (Smith *et al.*, 2008). The relevance of this problem is related to the possible serious consequences it can have on the cybervictims and cyberbullies.

From the traditional bullying literature, we know that bullying others has been associated with school problems (Rigby, 1997) and externalizing behaviours (i.e. Kumpulainen and Räsänen, 2000; Menesini *et al.*, 2009). Being victimized has been associated with the development of general psychological distress, increased levels of depression, anxiety and psychosomatic symptoms (O'Moore and Kirkham, 2001; Kochenderfer-Ladd and Skinner, 2002). Similar patterns of psychological health symptoms have been reported also for cyberbullying and cybervictimization (Ybarra and Mitchell, 2004; Mitchell *et al.*, 2007; Menesini *et al.*, in press). Some researchers have suggested that the cybercontext might produce even more serious consequences than the traditional bullying context, because of its characteristics (Kowalski and Limber, 2007; Gradinger *et al.*, 2009), such as spreading information to a wide audience, hiding responsibility through anonymity, the more permanent nature of written words and the ability to reach the target at any time and in any place (Campbell, 2007).

The need to intervene against cyberbullying is clear and urgent. To our knowledge, although there are several recommendations on how to prevent and reduce cyberbullying, no studies have as yet reported empirical data on the efficacy of the intervention (Willard 2007; Diamandorus, *et al.*, 2008).

Following the traditional bullying literature, several intervention programs have been developed and evaluated (Ttofi *et al.*, 2008; Smith *et al.*, 2004). On average, bullying was reduced by 20–23 per cent in experimental schools compared with control schools. Some of the most important programme components associated with a decrease in bullying and being bullied are: parent training, improved playground supervision, disciplinary methods, school conferences, videos, working with peers and cooperative group work (Farrington and Ttofi, 2009). An approach based on peer involvement and cooperation is developed through peer-led programmes, consisting of peer education and peer support actions (Shiner, 1999; Cowie and Wallace, 2000). Both models are based on the assumption that peers learn from each other, have significant influence on each other, and that norms and behaviours are most likely to change when liked and trusted group members take the lead in change (Shiner, 1999; Turner and Shepherd, 1999). Peer education and peer support have been used extensively in health-promoting projects and in particular to prevent and reduce bullying and conflict in schools (Cowie and Wallace, 2000; Campbell, 2004; Houlston and Smith, 2009; Maticka-Tyndale and Penwell Barnett, 2010).

Studies suggested that peer support systems and peer education can provide benefits for users of the scheme, peer supporters and schools in general (Naylor and Cowie, 1999; Cowie *et al.*, 2002; Maticka-Tyndale and Penwell Barnett, 2010). A significant decrease in bullying behaviours and in pro-bullying attitudes was found in different studies evaluating peer-led programs (Menesini *et al.*, 2003; Ellis *et al.*, 2005).

Overall, these findings suggest that both general peer-led interventions and specific peer support schemes may provide some beneficial outcomes, through trying to enhance students' responsibility for this problem and their own active involvement in prevention projects.

Cyberbullying by definition implies an active and predominant role by young people: because it is a peer problem; because of the digital divide between the old and new generations in the use of ICT; and because peers themselves are often the actors and the bystanders. Young people can be informants and contributors to our understanding of cyberbullying and can be effectively involved in designing and implementing interventions to prevent and reduce this problem among a community of peers (Spears *et al.*, 2009).

Starting from these considerations we assumed peer education and peer support to be suitable schemes to be implemented among young people to prevent and reduce cyberbullying. Developing a new scheme for cyberbullying means selecting the community, making a needs analysis of potential and risk factors, trying to adapt the model at school level, taking into consideration both face-to-face and virtual interaction, and designing roles and behaviours that could be supported in real and virtual life interaction. The aim of this chapter is to describe this peer-led scheme against cyberbullying and to evaluate the results in experimental schools.

The intervention program in Lucca: noncadiamointrappola (let's not fall into a trap)

The project commenced two years prior, with a first pilot project making use of a peer education scheme in face-to-face interactions. During the second year, two schools were involved in designing and developing a webpage and promoting some lines of virtual discussion and support through it. The web page name was 'Noncadiamointrappola' which means 'Let's not fall into a trap'. During the third year, six schools took part in the project as a project group and two schools as a control group.

During this period (2009–10), the project was carried out through the following stages:

1. Initial evaluation and needs analysis.
2. Launch of the project and awareness developing. Presentation of the project to the participating schools and classes, trying to raise awareness and communication on issues related to cyberbullying and bullying.
3. Selection of four on-line peer educators and four face-to-face peer educators in each participant class.
4. One day training for peer educators (8 hours).
5. Intervention of online educators in the forum NONCADIAMOINTRAPPOLA. We created a rotation schedule whereby all educators worked for a period of two weeks as forum moderators.
6. At the same time the intervention led by face-to-face educators was begun. In particular, face-to-face educators followed three lines of work: (a) preparation of a meeting run by the educators themselves with a school class who had not participated in the previous steps; the goal of this activity was to present the training and discuss cyberbullying issues with the class; (b) preparation of a meeting with the institutions in which educators would express their ideas in relation to emerging problems among young people in the city; and (c) preparation of the fourth episode of a TV program created by a local TV station where peer educators discussed bullying and cyberbullying with an expert.
7. Final evaluation: the questionnaire was re-administered to evaluate the final situation.

The present study

The present study aims to evaluate the efficacy of a peer-led intervention programme by examining the effects on cyberbullying and cybervictimization behaviours and on strategies used to prevent and to combat the problem through the longitudinal comparison of three participant groups: the control group (students who did not receive any kind of intervention); the experimental group 1 called 'awareness' (students who received only an intervention based on the awareness of cyberbullying); and the experimental group 2 called 'peer educators' (students who took part in the peer education training and actively worked in the real or virtual community).

Method

Participants and procedure

Participants in the study were 349 adolescents (62 per cent females), enrolled in 9th to 13th grades from eight high schools in Tuscany, Italy. The age of participants ranged from 14 to 20 years (mean age: 16.47; SD = 1.29). With regard to the high school typology, 37.4 per cent of students attended Lyceum high schools (this type of school comprises grammar and high schools), 34 per cent students attended Technical Institutes (part for industry and part for tourism), and 28.6 per cent students attended Vocational Schools (schools preparing for technical work). The schools were selected using a self-selection inclusion process and the classes were selected by the school staff. Self-report questionnaires were administered in class during school time by trained researchers. Consent procedure for research consisted of an approval by the schools and consent by the parents. Due to the prolonged collaboration with these school districts, 100 per cent of the families agreed to their children's participation in the research.

Overall, 236 adolescents completed the data at T1 and T2, thus we included only these students in the following analysis.

Measures

Self-reported cyberbullying and cybervictimization

We used two different measurement strategies in order to analyse the involvement in cyberbullying and cybervictimization: the first consisted of key questions and the second of Likert scales. With regard to the first strategy, after a definition of cyberbullying, the perception of involvement in cyberbullying and in cybervictimization was measured by two general key questions asking whether the adolescents had taken part in cyberbullying others or have been cyberbullied in the past couple of months (5-point scales: never, only once or twice; 2 or 3 times a month; about once a week; several times a week) (Smith *et al.*, 2008).

The second measurement strategy used two scales: one for perpetration and the second for victimization. A revised version of the Cyberbullying Scales (CS: Menesini *et al.*, in press) was used for this purpose. Each scale consisted of 18 items, asking how often adolescents had experienced several behaviours in the past couple of months. Each item was evaluated along a 5-point scale from never to several times a week. Reliability coefficients showed acceptable values: for cyberbullying, alphas were .81 at T1 and .75 at T2 and for cybervictimization, alphas were .72 at T1 and .83 at T2. A mean score was used for the following analysis.

Strategies to prevent or to intervene against cyberbullying

A multiple-choice question was used in order to evaluate the most effective ways to prevent and to combat cyberbullying. Participants were asked to answer with a maximum of three choices with several possibilities (see Table 12.1).

Results

Self-reported cyberbullying: cyberbullying scale

A mixed model ANOVA was carried out in order to evaluate the effect of time on cyberbullying across the three groups, considering also the effect of gender. Results showed a significant effect of time ($F(1,228) = 7.64$; $p < .001$), a significant interaction for time*group ($F(2,228) = 3.408$; $p < .05$) and a significant interaction for time*group*gender ($F(2,288) = 3.039$; $p < .05$). Figures 12.1 and 12.2 represent the change from T1 to T2 across the three groups. Looking at the main effect of time, the mean for cyberbullying decreases significantly from T1 to T2. However, interaction effects underline

Table 12.1 Strategies used to prevent and to reduce cyberbullying reported in T1 and T2 by the control group, awareness group and peer educators' group (%)

	Control Group		*Awareness Group*		*Peer Educators*	
	T1	*T2*	*T1*	*T2*	*T1*	*T2*
Talking with other adolescents about the internet risks	48	52	44	48	52	64
Blocking MSN	16	20	9	9	6	19
Denouncing or reporting to the authorities	66	66	58	64	65	67
Doing the same as cyberbullies	0	2	3	3	6	6
Ignoring	30	30	38	36	32	27
Changing e-mail or phone number	32	40	29	39	33	43
Trying to avoid risk situations	50	50	38	40	51	69

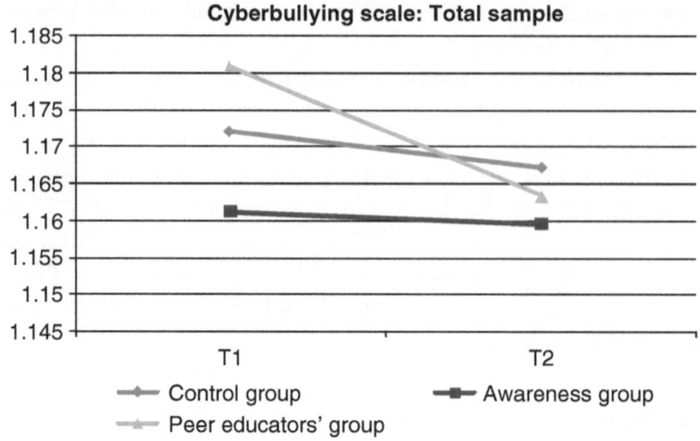

Figure 12.1 Cyberbullying means from T1 to T2 in control group, awareness group and peer educators' group

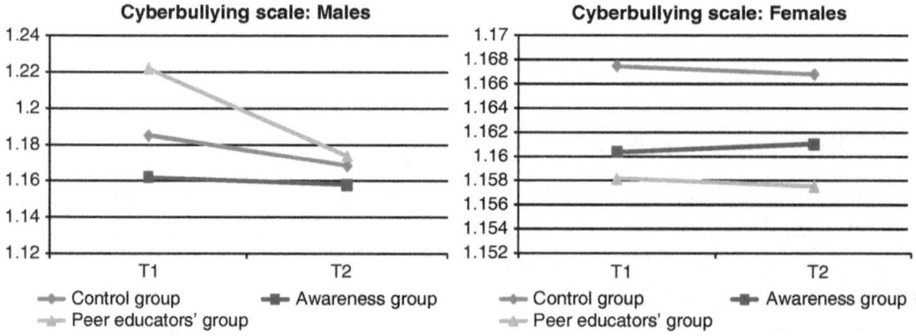

Figure 12.2 Cyberbullying means from T1 to T2 in control group, awareness group and peer educators' group differentiated by gender

that this decrease is not the same across groups and across gender. In order to statistically evaluate the effect of time across these groups, we calculated the significance of change within each group: results showed that only peer educators and in particular male peer educators showed a significant decrease across time (respectively: $F(1, 63) = 4.277$; $p < .05$; $F(1, 21) = 5.251$; $p < .05$). No other significant effect was found.

Self-reported cyberbullying: cyberbullying key question

To make the results comparable with traditional bullying intervention studies (e.g. Farrington and Ttofi, 2009), we investigated the extent to which the programme reduced the prevalence of cyberbullying using the key question ('Have you taken part in cyberbullying someone?'). To this end, we categorized students into cyberbullies or not involved by dichotomizing the self-reported cyberbullying responses, using once or twice as the cut-off point. As we can

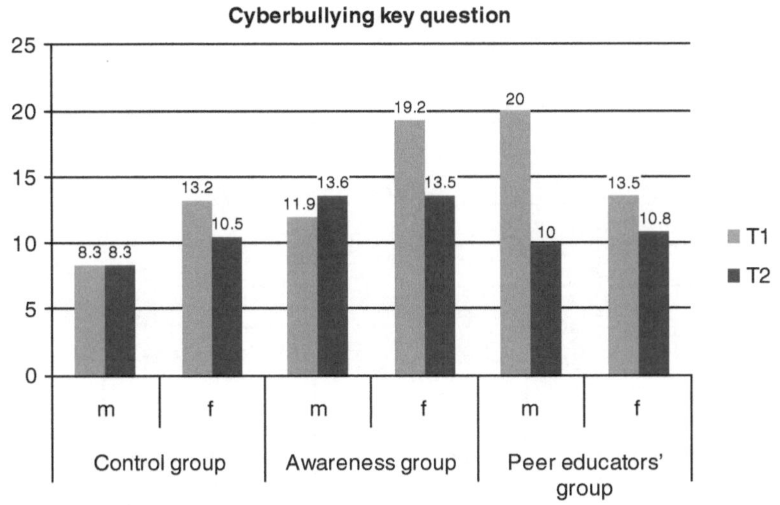

Figure 12.3 Cyberbullying key question: prevalence in the total sample

see in Figure 12.3, in the control group, males reported a stability from T1 to T2 and females reported a reduction of 20 per cent. In the awareness group, males reported a slight increase of 10 per cent from T1 to T2 and females reported a reduction of 30 per cent. Finally, in the peer educators' group, males reported a consistent reduction of 50 per cent from T1 to T2 and females reported a reduction of 20 per cent.

Self-reported cybervictimization: cybervictimization scale

A mixed model ANOVA was carried out in order to evaluate the effect of time on cybervictimization across the three groups, considering also the effect of gender. Results showed no significant main effect (time: $F(1, 230) = .015$; $p = .903$) and interaction effects (time*group: $F(2, 230) = .763$; $p = .467$; time*group*gender: $F(2, 230) = .102$; $p = .903$). Overall, results underlined a stability across time of cybervictimization for each of the three groups and for both genders.

Self-reported cybervictimization: cybervictimization key question

The same strategy used for the cyberbullying key question was used for cybervictimization. As we can see in Figure 12.4, males reported being non-involved in any kind of cybervictimization at T1 and T2 for both control group and peer educators' group. As for the awareness group, males showed a stability of involvement across time, of around 3 per cent. For females, the data are more differentiated: in the control group we have a consistent reduction of 66 per cent, in the awareness group a stability of

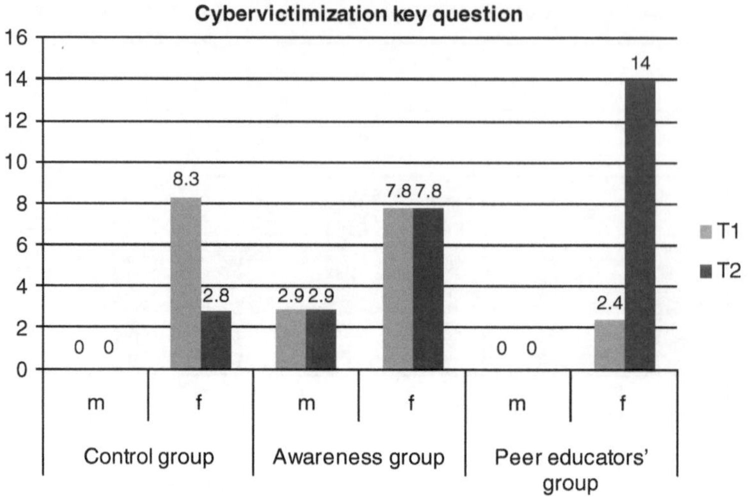

Figure 12.4 Cybervictimization item: prevalence in the total sample

involvement across time, at around 8 per cent, and for the peer educators' group results showed a consistent increase from 2 per cent at T1 to 14 per cent at T2.

Strategies to prevent or to combat cyberbullying

The seven strategies in T1 and T2 were compared across the three groups (see Table 12.1) in relation to how much students believed in their effectiveness. As can be seen in Table 12.1, peer educators reported an increase in the following strategies: talking with other adolescents about the internet risks, blocking MSN, changing mail or phone number, trying to avoid risk situations. On the other hand, the control group and awareness group showed a considerable increase only for changing e-mail or phone number.

Discussion and conclusions

Overall, the present study underlined that a peer-led intervention and peer scheme models can be a suitable approach to be implemented among young people to prevent and reduce cyberbullying. NONCADIAMOINTRAPPOLA, this project, showed positive effects in reducing cyberbullying mainly among peer educators, the students involved in the complete training, but not among the awareness group. In fact, especially this former group of students significantly decreased their level of cyberbullying and were more aware of different coping strategies that can protect possible victims. The results related to the cyberbullying behaviours highlighted a considerable decrease for peer educators taking into account both types of measures, the cyberbullying scale and cyberbullying key questions.

In relation to the coping strategies, an important and relevant finding concerned the role of proactive coping strategies, such as trying to avoid risk situations and inform other students about possible risk situations. Besides coping strategies that students can adopt, it is notable that the strategy 'denouncing or reporting to the authorities' is high for both the experimental and the control group at time 1 and remains high also at time 2. This finding tells us that young students see police or school authorities as a specific support to solve some of the most serious problems they might encounter.

A higher awareness level had been expected for the case of peer educators as compared to other groups. This was supported by the data confirming that taking the role of supporters and educators offers students a better opportunity to empower their skills and competence. From a theoretical point of view, the role theory suggests that taking a role can be an important process to enhance and reinforce skills and abilities of development (Sarbin, 1976; Bronfenbrenner, 1992). Other theories to explain the positive results of peer education and peer scheme models are: the theory of planned action, which stresses the role of aims, scope and behavioural control in promoting new actions and behaviours (Ajzen, 1991) and the theory of self-efficacy which underlines how behaving in a positive way can help to improve one's self-efficacy and consequently one's ability to fulfil specific tasks and situations (Bandura, 1997).

The project was less effective for the classes and for the rest of the participants defined as the 'awareness group'. This can be related to the type of intervention we carried out with these two groups: the peer educators' group worked more intensively through the training and other tasks they were asked to carry out. By comparison, the awareness group implemented a more extensive model but did not focus on the empowerment of social skills. This result highlights that an intervention based on explanations and lessons carried out by other persons, even if they are peers, is not sufficient to change the adolescents' behaviours. The opposite conclusion can be drawn in the case of high and personal involvement of peer educators.

However, the evaluation of the intervention underlined also that the project had some limitations for the following reasons: (1) the effectiveness of the intervention was limited to males; (2) the effectiveness of the intervention was limited to perpetrated behaviours. In relation to the first point, we need to point out that on average boys were more involved in cyberbullying than girls at time 1 evaluation, and particularly in the peer educators' groups: female peer educators reported a very low level of involvement as compared to males. Besides, males are often more impulsive and reactive in relation to others' provocations (Moffitt *et al.*, 2001). Therefore, an intervention focused on promoting awareness and positive behaviours can be more effective for boys than for girls.

In relation to the second point, the lack of efficacy for victims is a weak point of the project and needs further investigation in order to understand the possible causes. Some authors have underlined that victims need longer periods of intervention to change their perception of the situation (Naylor and Cowie, 1999). Any future efforts by this project should therefore address the victims more directly and try to give them support and protection. Probably a greater focus on peer support than on peer education can assist.

Implications

The study underlines the importance of prevention and intervention to reduce risks on the web and particularly for cyberbullying. It also demonstrates that effective interventions should be run by students themselves and should involve young people in a positive way, as observers and participants. In our case, new lines of interventions were developed and designed by children themselves with the help and the supervision of adults. A peer-to-peer communication on which strategies are safe, to help others to cope with the problems and in general to promote positive communication and relationships within the virtual community resulted in a promising approach.

Finally, other ways to improve the project can be proposed: (1) given the positive effects found with the peer educators, the next step could consist of evaluating whether this intervention can have similar effects also on the target peers (i.e. those students who received the intervention); (2) the need to integrate cyberbullying prevention with the school policy is an important issue that the Italian school system should take into consideration in the future steps (Menesini *et al.*, in press, b).

References

Ajzen, I. (1991) 'Theory of planned behavior', *Organizational Behavior and Human Decision Processes*, 50: 179–211.

Bandura, A. (1997) *Self-Efficacy: The Exercise of Control*, New York: Freeman.

Bronfenbrenner, U. (1992) 'Ecological system theory', in R. Vasta (ed.) *Six Theories of Child Development*. London: Jessica Kingsley.

Campbell, C. (2004) 'Creating environments that support peer education: experiences from HIV/AIDS-prevention in South Africa', *Health Education*, 104(4): 197–200.

Campbell, M.A. (2007) 'Cyber bullying and young people: treatment principles not simplistic advice', available at: www.scientist-practitioner.com, paper of the week, 23 February.

Cowie, H. and Wallace, P. (2000) *Peer Support in Action: From Bystanding to Standing By*, London: Sage.

Cowie, H., Naylor, P., Talamelli, L., Chauhan, P. and Smith, P.K. (2002) 'Knowledge, use of and attitudes towards peer support: a two year follow-up to the Prince's Trust survey', *Journal of Adolescence*, 25(5): 453–67.

Diamandorus, T., Downs, E. and Jenkins, S.J. (2008) 'The role of school psychologists in the assessment, prevention, and intervention of cyberbullying', *Psychology in the Schools*, 45(8): 693–704.

Ellis, L.A., Marsh, H.W. and Craven, R.G. (2005) 'Navigating the transition to adolescence and secondary school: a critical evaluation of the impact of peer support', in H.W. Marsh,

R.G. Craven, and D.M. McInerney (eds) *New Frontiers for Self Research*, Greenwich, CT: Information Age Publishing, pp. 329–55.

Farrington, D.P. and Ttofi, M.M. (2009) 'How to reduce school bullying', *Victims and Offenders*, 4(4): 321–6.

Gradinger, P., Strohmeier, D. and Spiel, C. (2009) 'Traditional bullying and cyberbullying: identification of risk groups for adjustment problems', *Zeitschrift für Psychologie*, 217(4): 205–13.

Houlston, C. and Smith, P.K. (2009) 'The impact of a peer counselling scheme to address bullying in an all-girl London secondary school: a short-term longitudinal study', *British Journal of Educational Psychology*, 79: 69–86.

Kochenderfer-Ladd, B. and Skinner, K. (2002) 'Children's coping strategies: moderators of the effects of peer victimization?' *Developmental Psychology*, 38: 267–78.

Kowalski, R.M. and Limber, S.P. (2007) 'Electronic bullying among middle school students', *Journal of Adolescent Health*, 41: S22–S30.

Kumpulainen, K. and Räsänen, E. (2000) 'Children involved in bullying at elementary school age: their psychiatric symptoms and deviance in adolescence: an epidemiological sample', *Child Abuse & Neglect*, 24: 1567–77.

Maticka-Tyndale, E. and Barnett, J.P. (2010) 'Peer-led interventions to reduce HIV risk of youth: a review', *Evaluation and Program Planning*, 33: 98–112.

Menesini, E., Codecasa, E., Benelli, B. and Cowie, H. (2003) 'Enhancing children's responsability to take action against bullying: evaluation of a befriending intervention in Italian middle schools', *Aggressive Behavior*, 29: 1–14.

Menesini, E. and Nocentini, A. (2009) 'Cyberbullying definition and measurement: some critical considerations', *Zeitschrift für Psychologie*, 217(4): 230–2.

Menesini, E., Modena, M. and Tani, F. (2009) 'Bullying and victimization in adolescence. concurrent and stable roles and psychological health symptoms', *The Journal of Genetic Psychology*, 170(2): 115–33.

Menesini, E., Calussi, P. and Nocentini, A. (in press (a)) 'Cyberbullying and traditional bullying: unique, additive and synergistic effects on psychological health symptoms', in Q. Li, D. Cross, and P. Smith (eds) *Bully Goes to the Cyber Playground: Research of Cyberbullying from an International Perspective*, Chichester: Wiley-Blackwell.

Menesini, E., Nocentini, A. and Calussi, P. (in press (b)) The measurement of cyberbullying: dimensional structure and relative item severity and discrimination', *Cyberpsychology, Behavior, and Social Networking*.

Mitchell, K.J., Ybarra, M.L. and Finkelhor, D. (2007) 'The relative importance of online victimization in understanding depression, delinquency, and substance use', *Child Maltreatment*, 12: 314–24.

Moffitt, T.E., Caspi, A., Rutter, M. and Silva, P.A. (2001) *Sex Differences in Antisocial Behaviour: Conduct Disorder, Delinquency and Violence in the Dunedin Longitudinal Study*, Cambridge: Cambridge University Press.

Naylor, P. and Cowie, H. (1999) 'The effectiveness of peer support systems in challenging school bullying: the perspectives and the experiences of teachers and pupils', *Journal of Adolescence*, 22(4): 467–80.

O'Moore, M. and Kirkham, C. (2001) 'Self-esteem and its relationship to bullying behaviour', *Aggressive Behavior*, 27: 269–83.

Rigby, K. (1997) *Bullying in Schools, and What to Do About It*. London: Jessica Kingsley.

Sarbin, T.R. (1976) 'Cross-age tutoring and social identity', in V.L.Allen (ed.) *Children as Teachers: Theory and Research on Tutoring*, New York: Academic Press, pp. 488–567.

Shiner, M. (1999) 'Defining peer education', *Journal of Adolescence*, 22(4): 555–66.

Smith, P.K., Pepler, D. and Rigby, K. (eds) (2004) *Bullying in Schools: How Successful Can Interventions Be?* Cambridge: Cambridge University Press.

Smith, P.K., Mahdavi, J., Carvalho, M., Fisher, S., Russell, S. and Tippett, N. (2008) 'Cyberbullying: its nature and impact in secondary school pupils', *The Journal of Child Psychology and Psychiatry*, 49(4): 376–85.

Spears, B., Slee, P.T., Owens, L.D., and Johnson, B.E. (2009) 'Behind the scenes and screens: insights into the human dimension of covert and cyber bullying', *Zeitschrift für Psychologie*, 217(4): 189–96.

Ttofi, M., Farrington, D. and Baldry, A.C. (2008) *Effectiveness of Programmes to Reduce School Bullying: A Systematic Review*, report prepared for The Swedish National Council for Crime Prevention.

Ttofi, M.M. and Farrington, D.P. (2009) What works in preventing bullying? Effective elements of anti-bullying programmes. *Journal of Aggression, Conflict and Peace Research*, 1, 13–24.

Turner, G. and Shepherd, J. (1999) 'A method in search of a theory: peer education and health promotion', *Health Education Research*, 14(2): 235–47.

Willard, N. (2007) *Educator's Guide to Cyberbullying and Cyberthreats:* Center for Safe and Responsible Use of the Internet, available at: http://www.cyberbully.org/docs/cbctparents.pdf (accessed 14 July 2010).

Ybarra, M.L. and Mitchell, K.J. (2004) 'On line aggressor/targets, aggressor and targets: a comparison of associated youth characteristics', *Journal of Child Psychology and Psychiatry*, 45(7): 1308–16.

Chapter 13

Cybermentoring

Jessica von Kaenel-Flatt and Thaddaeus Douglas

'How RU today?' That's what CyberMentors want to know, in an innovative model of peer mentoring that takes place both on and offline. Launched in March 2009 by the national charity Beatbullying (www.beatbullying.org), CyberMentors was conceived as a response to bullying – including but not limited to cyberbullying – that directly empowers young people to help their peers through a medium that is natural and comfortable for them.

How does it work?

Young people can be trained as CyberMentors from the age of 11 to 16. The majority are recruited from participating schools, either nominated by a member of school staff or self-selected, but many graduate through the mass trainings that are held in communities across the UK. The recruitment process for participating schools is two-way: schools interested in taking on the programme can contact Beatbullying to arrange training sessions for their pupils, and Beatbullying approaches local authorities (the regional councils with responsibility for overseeing education in each area of the country) and schools are recruited through them. Sometimes it is the young people who use the site who get their schools involved, if their schools do not have the programme already.

During the two-day training, the young people are equipped with the skills to listen empathetically to their peers, give appropriate advice, use the functions of the site and, just as importantly, understand when an issue goes beyond what a peer mentor can deal with and needs to be passed on to a trained adult. CyberMentors support their peers online, through e-mail-type messages and real-time private chats on the website, for a ten-minute session at a time. Some schools allow mentors to access the site during breaks and lunchtimes – this is up to the individual school – but obviously not during lessons; mentors use their own free time to come online and offer support. CyberMentors also set up offline peer mentoring schemes, adding to existing schools-based support strategies for children. For everyone's safety, it is strictly against the rules for anyone to offer mentoring unless they have completed and graduated from the official CyberMentors' training; however, there is a high degree of unofficial support among the community of users in general chat. As in any good community, the members look out for each other.

The site employs a hierarchy of support: CyberMentors are trained to listen to and advise their peers of roughly the same age about the kinds of problems that many young people encounter at school and at home, mainly bullying but also other issues that are problematic but not matters of severe and immediate danger. Senior CyberMentors are young people aged 16–25, many of whom are at college, university or work, who are trained to support their younger counterparts with mentoring issues but also to mentor in their own right. As young adults they often have the maturity and experience to take on more difficult or complex issues, and to engage with older users who might well be facing different challenges from younger ones. However, no mentor of any age is expected or trained to deal with very serious and potentially life-threatening issues such as abuse, self-harm, eating disorders, substance abuse, mental illness or the intention of suicide. At this level, Beatbullying employs BACP (British Association for Counselling and Psychotherapy) accredited professional counsellors who are available every day from 8am to 2am. Their role is to engage deeply with the young people who have chosen to enter counselling, within the framework of a clearly delineated therapeutic relationship: the counsellor provides an explanation of what counselling is, what the user can expect, the limits of confidentiality and a structure of ground rules. If the user understands and agrees, they are contracted into counselling for an agreed period. During this time they are asked to maintain necessary boundaries for their own emotional health, such as discussing presenting issues in depth only with their own assigned counsellor. This service, which is part of CyberMentors and operates on the same site, is free at the point of delivery to all users who request it.

The types of issues which fall in and outside a mentor's scope are clearly delineated and serious disclosures are immediately passed on. The site has a section which clarifies to users the limits of confidentiality: basically, everything a young person says to a mentor is between the two of them unless it is felt that someone is in serious or immediate danger or their situation is beyond the scope of a peer mentor to deal with. If this is the case, then they will try to let the person know. However, it is generally recognized that there are times when concern for someone's safety has to override the right to privacy. Counsellors, moderators and members of staff act as contact points for mentors and other users who have concerns about someone on the site. Beatbullying has a dedicated Child Protection team that liaises with social and emergency services to step in, in real time, in the most urgent cases.

Usually, people who want to use the site will register. By completing a few basic questions that help with demographic monitoring and tracking but are not visible to other users of the site, a young person becomes a

Registered User with a unique username, which can be anything they choose as long as it is not offensive and does not include identifying details. They also have a widget, the sweet and colourful CyberMentors trademark avatar of a computer screen with a face. Once registered, the user can join the general chat room; send and receive private messages; open private chats with other users, mentors and staff; build up and customize a profile page about themselves with pictures, wallpaper and links to approved sites; and submit blogs for publication. This is available to any young person under 18 whether or not they live in the British Isles – we have had registered users resident in the USA, Australia, Vietnam and Japan, and are looking at ways we might possibly develop our work internationally. However, for various reasons not everyone who comes to the site for help wants to register. If this is the case, the young person can either leave a message on an online forum which will be picked up by a CyberMentor or leave a message for a counsellor, and need not contain any more information than the person's contact details.

CyberMentors (www.cybermentors.org.uk) is not exclusively a site for young people experiencing problems. While its primary function is to offer help and support, it is also designed to allow children and young people to chat and build online friendships in a safe environment. The site employs a multi-tiered system of moderation for language and content: any potential interaction in the open chat room or through private messaging is first filtered through the cutting-edge 'Crisp' moderation software which automatically blocks words and phrases identified as being inappropriate, which includes swearing, insults and sexual references as well as requests for private information such as address or date of birth. In addition to this, Beatbullying also employs human moderators, all of whom have been trained in person by Beatbullying staff. This assures the site users of a friendly, trustworthy adult presence on the site to engage in chat, keep a look-out for any problems and act as a point of contact for any questions or concerns. Everyone employed by Beatbullying, whether or not they have direct contact with children and young people, is required to undergo an enhanced Criminal Records Bureau (CRB) check repeated every three years from the beginning of employment. All staff and moderators in the chat room are fully visible to users, and join in the conversation with users and mentors. In this way, relationships of trust are built and the adults are not seen as being there only to intervene when there is trouble, although this is part of the job too. The site has a set of rules and a system of consequences for infractions, usually two warnings followed by a 7-day ban from general chat, although users can still access mentoring and counselling during this time.

The site rules are stated in as clear and user-friendly a way as possible and every user is asked to read and make sure they understand them before signing up. They were developed for the safety and comfort of everyone who uses the site and the reason for each of them is explained in a way that young people will hopefully understand. Users and mentors are not allowed to ask for or give certain kinds of personal information that could be used to identify them, such as real name, location, age or school; post links to their profiles on social networking sites or to any website that contravenes the ethos of CyberMentors; or post photographs of themselves or people they know on their profiles. They are asked to respect the site and the people who use it by observing basic courtesy in human relationships: no bullying, swearing, harassment, intrusive questions or trying to find romantic or sexual relationships through the site. In the chat room there are some points of etiquette that users are asked to observe: using a consistent font size; no spamming, such as constant repetitions of meaningless words; phrases or letters; no jumping in and out of the room; most importantly, users are asked to remember that general chat is for distraction, not for discussing serious or disturbing issues. Users who post comments or status messages alluding to serious problems such as self-harm or suicide, or using a lot of sad-face emoticons, are contacted by a moderator or counsellor and offered support.

The CyberMentors site is strictly for people up to the age of 18, the age of legal adulthood in the UK. Very occasional exceptions have been made where an existing user, under 18 at the time of signing up, is identified as needing extra help in making the transition from the site to other forms of support. In those cases the young person would be guided by professional counsellors to move gradually into specific young adult provision. However, this transition gap is now being bridged by Beatbullying's development of FutureYou, a dedicated site for 14–25-year-olds which provides advice and support on employment, training and education. It is possible for interested parents or professionals to take a guided tour of the site by appointment, and there is a consistent adult presence of staff, moderators, counsellors and Senior CyberMentors, but apart from that, the site is for young people only. This is not just for reasons of safety, important as these are. It is also a matter of ownership. CyberMentors exists to empower young people to help each other; as far as possible the site should be *for them* in every sense.

Why does it work?

The power of young people to harm each other through communications technology is well attested. According to research by Wolak (2006), cited in Palfrey (2008), one-third of reported cyber-aggression was 'distressing', with up to 76 per cent of targets reporting upset, fear or embarrassment and 35 per cent experiencing other effects such as avoiding the internet, feeling jumpy or irritable and losing interest in things. Patchin and Hinduja (2006) (http://www.cyberbullying.us/research.php) find 54 per cent of targets experiencing emotions such as frustration, anger and sadness, and Ybarra and Mitchell (2004) report that bullies, victims and bully/victims (those who are at different times both victim and aggressor) are more likely than others to exhibit serious depressive symptoms. Figures from the CyberMentors website (www.cybermentors.org) indicate that experiencing bulling victimization is a significant factor in driving young people to physically self-harm; Beatbullying's latest research shows that as many as 44 per cent of all child suicides in the UK may well be down to bullying. Less immediately drastic but still highly concerning, research conducted by Beatbullying in 2006 (http://www.beatbullying.org/dox/resources/reports.html) found that 36 per cent of all school truancy was directly linked to bullying; missing days of school has a direct effect on future educational prospects, employment and earning potential in adulthood as well as the less quantifiable aspects of well-being such as self-esteem and sense of achievement.

The evidence shows, then, that the words and actions of young people have a huge impact on their peers. The effectiveness of CyberMentors lies in harnessing this power for good. Many features of cyberspace that are so effective in threatening and disempowering victims of bullying – the potential for relentless, round-the-clock contact; the perceived anonymity of faceless, unknown aggressors; the potential for instantaneous dissemination of material to a limitless audience; the fact that with the right technology it can be accessed from anywhere – are also exactly the qualities which, turned on their heads, can be most helpful to a young person seeking support. There is an online presence of mentors, moderators and counsellors throughout much of the day and night; any number of them can be reached at the touch of a button; any young person with access to the internet anywhere can use it, however limited and sporadic their access may be; and perhaps most importantly to many users, the partial anonymity afforded by the online nature of the service makes it possible to disclose as much or as little as the person wishes. In interviews with users of the site conducted in March 2010, this aspect was identified as highly valued, particularly by users who would find face-to-face mentoring overwhelming or to whom it is simply not available.

Evaluation of the programme is ongoing. Our research and evaluation partners are the University of Sussex, New Philanthropy Capital, and Beatbullying has also been evaluated by Goldsmiths, University of London.

Added to this, Beatbullying employs a Research and Development Team dedicated to research and evaluation, thus highlighting our commitment to measure the impact and inform the development of our work.

We adopt a combination of research methodologies to evaluate the effectiveness of our interventions. The schools-and-communities-based mentoring model has seven levels of questionnaires introduced to the young people and the professionals working with the young people over a 12-month period. Additional strategies are employed, including in-depth semi-structured interviews, focus groups, self-reporting questionnaires and survey assessment tools, documenting, and video/audio evaluation (both online and offline).

In schools where the programme has been implemented, staff consistently report a reduction in bullying – including cyberbullying – by an average of

43 per cent. In addition, reporting of bullying incidents including cyberbullying goes up by an average of 60 per cent. Adjusted figures over a 12-month period are 40 and 33 per cent respectively.

Who uses CyberMentors?

It is generally recognized that those who bully tend to pick on targets who appear in some way 'different' from the majority, or the average. This may be for any reason, including some that appear counter-intuitive such as being popular or having a talent, but there are some groups who are officially identified as being particularly vulnerable: those with a disability or SEN (Special Educational Need), those from certain minority ethnic backgrounds, those receiving free school meals and those living in foster care or children's homes.

Beatbullying's own large-scale and generally representative research, a study of cyberbullying among 2094 secondary-age pupils in 20 English comprehensive schools whose findings are written up in the *Virtual Violence* report (Cross *et al.*, 2009), has found that the *overall* instance of cyberbullying – encompassing every experience from an isolated incident to concerted long-term aggression – does not differ markedly by social category. However, for children in certain officially defined 'vulnerable' groups, the incidence of being *persistently* bullied, where bullying is ongoing over a significant period of time: months to years, via the internet or mobile phone technology is significantly higher than the national average. Sixteen per cent of those with Special Educational Needs, 13 per cent of those receiving free school meals and 24 per cent (almost a quarter) of those identifying ethnically as 'White Other', i.e. white non-British ethnic groups such as Eastern European or those outside the cultural mainstream including Gypsy/Roma, reported being targets of persistent bullying. This compares with 9 per cent of children without official 'vulnerable' status. In addition, our findings show that twice as many girls as boys experience this sustained form of aggression. Crucially, in terms of resourcing interventions and targeting behaviour change campaigns, Beatbullying has worked in partnership with a wide range of statutory and non-statutory partners to target such groups and offer consistent and ongoing support.

Impact and reach

Since the launch of the site in March 2009, over 4000 young people have been trained as CyberMentors and Senior CyberMentors; 980,199 children and young people have accessed the website for help, support and information, and the homepage has been viewed over 9 million times.

'CyberMentors' is a multi-award winning programme, awarded for Best Communications Campaign in the 2010 Third Sector Excellence Awards; and Best Use of Digital Media at the Third Sector Excellence Awards 2009.

Interest in the programme has been registered from as far afield as Brazil, Italy and Japan. In 2010, CyberMentors won an award in the 'Making the Internet Safer' category of the Nominet Internet Awards.

Perhaps the best testimony, though, is not anything that can be encapsulated in a statistic but the young people who first came to the site as victims needing help and reappear a few months later as mentors themselves.

Although part of a mentor's role is giving advice when needed, this is a relatively small aspect of what they actually do. The really vital aspects of their work are building up the confidence and sense of self-worth that bullying takes away, being there as a friendly presence to listen and by doing so to reassure the other person that they are not alone. When a CyberMentor says, 'How RU today?' they really want to know. And for some young people, that might be the first person who ever has.

References

Beatbullying (2006) *Bullying and Truancy Report*, available at: http://www.beatbullying.org/dox/resources/reports.html.

Cross, E.-J., Richardson, B., Douglas, T., von Kaenel-Flatt, J., (2009) *Virtual Violence: Protecting Children from Cyberbullying*, available at: http://www.beatbullying.org/dox/resources/reports.html.

Palfrey, J., Boyd, D. and Sacco, D. (2008) *Enhancing Child Safety and Online Technologies: Final Report of the Internet Safety Technical Task Force to the Multi-State Working Group on Social Networking of State Attorneys General of the United States*: Dec. 31, Berkman Center for Internet and Society at Harvard University.

Ybarra, M.L. and Mitchell, K.J. (2004) *Online Aggressor/Targets, Aggressors and Targets: a Comparison of Associated Youth Characteristics*, Johns Hopkins Bloomberg School of Public Health, Center for Adolescent Health Promotion and Disease Prevention, Baltimore, MD /Crimes Against Children Research Center, University of New Hampshire, Durham, NH, USA. *Journal of Child Psychology and Psychiatry*, 45:7, available at: http://www.unh.edu/ccrc/pdf/jvq/CV75.pdf.

Chapter 14

Bullies' and victims' experiences of the anti-bullying game from the KiVa program

Elisa Poskiparta, Ari Kaukiainen, Virpi Pöyhönen and Christina Salmivalli

In October 2006, a commercial video game called 'Bully' (also known as Canis Canem Edit, 'Dog Eat Dog') was released in North America and the UK. It is no wonder that the game became highly controversial: due to its glorification of bullying behavior and sexual content it can be considered ethically precarious at least. The concerns raised by educators are justified. Bullying, defined as repeated aggressive behavior against a victim who cannot readily defend himself or herself (Olweus, 1994) is a common problem in schools worldwide (Smith *et al.*, 1999). As meta-analyses have confirmed that playing violent games leads to increases in aggression (Anderson and Bushman, 2001; Sherry, 2001) as well as decreases in prosocial behavior (Anderson and Bushman, 2001), it is not far-fetched to ask whether a game where youth are put into the role of an aggressive bully enhances such behaviours in real-life settings.

As children and youth are enthusiastic consumers of modern technologies and computer games anyway, it is important to find ways in which these tools can be used to enhance learning of positive things. With respect to bullying, the crucial question is whether information and communication technologies could be used to reduce bullying, to help children cope with bullying, or to behave constructively when witnessing it? ICT has been used to address the problem of bullying mostly by various anti-bullying internet platforms where the visitors are provided with information about bullying and guidelines to tackle it.

To our knowledge, FearNot! is the only anti-bullying game whose effects have so far been investigated by an intervention-control design (Sapouna *et al.*, 2010). The game is designed for children who are either victimized or at elevated risk for victimization, and it provides a safe virtual environment where children can try out strategies for dealing with bullying. In the evaluation study by Sapouna and colleagues, the FearNot! intervention lasted for three weeks and consisted of three sessions delivered to 8–9 year-old students in school classes in the UK and Germany. The intervention increased children's self-reported probability of escaping victimization. Unfortunately, the effect was only found among British, not German students and only one week, and didn't last more than four weeks after the intervention. Limited success is understandable as the intervention was short-term (three sessions, each lasting 30 minutes over a three-week period). The authors concluded that future applications will have to explore whether a virtual learning intervention is most suitable as an additional component within a wider antibullying curriculum.

KiVa anti-bullying programme

The recently developed KiVa anti-bullying program (Salmivalli *et al.*, 2010) includes both universal (e.g., student lessons) and indicated actions (e.g., school teams tackling the cases of bullying coming to attention) to reduce peer-to-peer bullying in schools. The universal actions involve three slightly different versions targeted at students in grades one, four, and seven. The program has a multifaceted theoretical background: it is based on research on participant roles in bullying (Salmivalli *et al.*, 1996) as well as on recent studies on the social standing of aggressive children in general (e.g., Rodkin, et al, 2006) and bullies in particular (Juvonen *et al.*, 2003). The program is predicated on the idea that bystanders often contribute to the emergence and maintenance of bullying by reinforcing the bully (Salmivalli *et al.*, 2011), which in turn provides bullies with the status and power that they seek (Sijtsema *et al.*, 2009; Salmivalli, 2010). It is thus assumed that a positive change in the behaviours of classmates can reduce the social rewards gained by bullies and consequently their motivation to bully. When we started to develop the KiVa program, the idea of utilizing the powerful learning media provided by ICT was present from the beginning. Consequently, the program includes, besides more traditional remedies to tackle bullying, an anti-bullying computer game. There are separate, developmentally appropriate versions for Grades 1, 4, and 7.

KiVa computer game for fourth-graders

In the context of the KiVa program, the purpose of the anti-bullying game is to motivate students and enhance their learning process. Although the game can be played independently from the KiVa program as well, it is closely connected to the twenty (ten double) student lessons of KiVa. The contents of the game are designed to match the topics brought up during student lessons and the game is typically played during and between these lessons according to a schedule recommended by program developers (see Table 14.1). During the first phase of the evaluation of KiVa, the current fourth-graders' version of the program was evaluated in grades four to six. Thus, what is now a fourth-graders' computer game was played by students in grades four, five, and six.

The setting of the KiVa game is a virtual school and the non-player characters, which are the animated characters the player does not control, are students in this school. In order to increase students' motivation to play the

Table 14.1 The contents of the KiVa student lessons and the corresponding topics in the KiVa computer game

Student lesson	*Level of the game*	*Module of the game*		
		I Know	*I Can*	*I Do*
1. Respect is for everybody 2. In a group 3. Recognize bullying	Level 1(October, during/after lesson 3)	1. Test yourself: Your ways to get to know people 2. Is this communication? 3. Ask the police! 4. What does KiVa stand for? 5. How to get friends? 6. Do you know the word? - Group, group pressure, role 7. Do you remember the KiVa rules 1-3?	1. Is this bullying or not? Two episodes in the lunchroom Two episodes in the playground One episode in the corridor One episode in the schoolyard	1. We treat everyone with respect 2. Together we make our class a comfortable place for everyone. 3. We don't bully others.
4. Hidden forms of bullying 5. Consequences of bullying	Level 2 (December, during/after lesson 5)	1. Test yourself: What do you think and how do you act if you hear people say nasty things about someone? 2. What are your ways to defend yourself if you have been bullied by cell phone or through the internet? 3. What are the consequences of bullying to a victim? 4. Do you know the word? - Hidden forms of bullying, fear, mobile phone bullying 5. Do you remember the KiVa rules 4-5?	1. Bullying from the perspective of the bully and the victim Letters from a bully and from a victim 2. How can I defend the victim? Two episodes in the playground One episode in the recess place	

Table 14.1 (continued)

Student lesson	*Level of the game*	*Module of the game*		
		I Know	*I Can*	*I Do*
6. Group involvement in bullying 7. Countering bullying as a group	Level 3 (February, during/after lesson 7)	1. What prevents me from defending the victim? 2. Identifying participant roles in bullying 3. How I can support and defend the victim? Which ways are easy/difficult for me? 4. Do you remember the KiVa rules 6-7?	Bullying from the perspectives of the bully and the victim. Letters from a bully and from a victim How can I defend the victim? Two episodes in the corridor One episode in the recess place	1. We don't bully others or reinforce the bully 2. We support and defend the victimized peers
8. What to do if I get bullied?	Level 4 (March, during/after lesson 8)	1. What are my ways to defend myself if someone bullies me? Which ways are easy/difficult for me? 2. What to do if I get bullied? - Can you identify assertiveness? - What do you do, if you have been bullied? - Why it is worthwhile to report bullying to adults? 3. Do you know the word? - Assertiveness, shame, telling 4. Do you remember the KiVa rule 8?	What to do if I get bullied? One episode in the recess place One episode in the lunchroom	1. We report bullying to adults
9. KiVa school - let's do it together! 10. How are we doing?	Level 5 (May, during/after lesson 10)		Exploring the situation and feelings of the virtual characters (former victims, bullies, etc.) at the end of the school year	

KiVa game, they are able to create their own player-character and influence its physical appearance (e.g., hair style, skin colour, clothes and other equipment such as mobile phone, brooches).

The KiVa game involves five levels and it is mainly supposed to be played at school during selected student lessons, always proceeding to the next level as soon as a particular lesson has been given in the classroom. The students can also play the game in their free time outside the school. The topics are designed to match the issues brought up in student lessons or theme days, and they proceed from more general topics, such as respectful communication and resisting group pressure, to bullying and its mechanisms and consequences. The game is started after the third lesson, the second level is played after the fifth lesson, and so forth, during the whole school year. Each level of the KiVa game includes three modules that have been named as I Know, I Can, and I Do. As displayed in Table 14.1, the topics of the ten double lessons are related to the issues brought up in the *I Know*, *I Can*, and *I Do* modules in each of five levels of the KiVa game.

In the *I Know* module, the students learn new *facts* about bullying but also examine what they have learnt during the lessons so far. They are asked questions about the contents of the lessons in game-like quizzes, and they can test themselves with respect to different characteristics (e.g., Can I resist group pressure? What are my best ways to support a victimized peer? See Table 14.1).

In the *I Can* module, the students practise the *skills* they have learnt during the student lessons. They (e.g. the characters they've created) move around in a virtual school and face different challenging situations in the playground, lunchroom, and school corridors. They make decisions regarding how to respond to these situations, what to say and what to do – and get feedback based on the choices they have made. For instance, if someone is being bullied during the recess, the player is first asked to reflect on his/her thoughts (What do you think when you see the situation?). Then the player faces different options to act in the situation (e.g. walking closer; going away; intervening in the situation; telling the teacher). After the player has chosen an option, the situation continues and the player is again asked to choose from different options to behave. In certain points of the narration the player has an opportunity to 'read the minds' of the other characters (i.e. the victim or the bystanders), seeing how they think and feel. Based on these cues, and on how the episode proceeds, the player can also change his/her behaviour and try out something different. A scene taking place in the virtual schoolyard in the *I Can* module is presented in Figure 14.1.

The third module, *I Do*, is designed to encourage the students to make use of their acquired knowledge and skills in real-life situations. This happens by asking them to report – at each level of the game – which ones of the KiVa rules (that were adopted during the lessons) they've succeeded in putting into practice (see Table 14.1).

Figure 14.1 KiVa screenshot

A study of students' perceptions of the KiVa computer game

The effectiveness of the KiVa anti-bullying program as a whole has already been evaluated in a large-scale randomized controlled trial (the effects on grades 4–6, see Kärnä *et al.*, 2011; the effects in grades 1–3 and 7–9, see Kärnä *et al.*, under review). The aim of the present study was to shed light on the potential benefits of utilizing new technologies, and especially e-learning methods as part of such a whole-school program. We examined students' use and experiences of the KiVa computer game, investigating the effects of sex, grade level, and bullying status (i.e., whether a student is a victim, bully, bully-victim, or not involved in bullying) on their experiences and perceptions.

The data for the present study were collected in the first phase of the evaluation of KiVa, when the program effects were tested in grade levels four to six (Kärnä *et al.*, 2011). Thus, the participants were from grades 4, 5, and 6 and approximately 10, 11, and 12 years of age, respectively. The data came from two different sources: (1) a large web-based questionnaire involved in the evaluation study of the KiVa program and filled in by the students before (Spring, 2007), in the middle (Winter, 2008) and at the end (Spring, 2008) of the evaluation year; and (2) a short questionnaire presented to students in the last level (level five) of the KiVa computer game. As each student logged into the large internet-based questionnaire and to KiVa game with the same password, we were able to combine the information from these two data sources.

From the web-based questionnaire data we utilized the post-test data which was collected around the same time (May 2008) when the students played the last level of the computer game and thus reported on their use and perceptions of the game. The post-test data were provided by 3553 students and the game data by 1319 students; both data were available from

1198 students. However, after excluding the few bully-victims (see below) from the analyses, the final sample utilized in the present study consisted of 1188 children.

From the large-scale evaluation data, we utilized children's self-reports of bullying others and being victimized, in order to classify them as victims, bullies, bully-victims, and non-involved. Students responded to the global items derived from the Olweus' Bully/Victim questionnaire (Olweus, 1996): 'How often have you bullied others at school in the last couple of months?' and 'How often have you been bullied at school in the last couple of months?' Students answered on a 5-point scale (0 = *Not at all*, 1 = *Once or twice*, 2 = *Two or three times a month*, 3 = *Every week*, 4 = *Several times a week*), and their classification to the four groups was based on the cut-point of 'two or three times a month' (Solberg and Olweus, 2003). In other words, victims (7.7 per cent of the sample, 8.4 per cent of those providing game data) were students who said they had been bullied by others at this frequency or more often, bullies (2.1 per cent of the sample, 2.5 per cent of those providing game data) told they had bullied others with this frequency or more often, and bully-victims (1.4 per cent of the sample, 0.8 per cent of those providing game data) were the ones who reported both bullying others *and* being bullied by others 2–3 times a month or more often. The rest of the students were classified as non-involved (88.8 per cent of the sample, 88.2 per cent of those providing game data). Because bully-victims were clearly under-represented among children who provided data on their use and perceptions of the computer game (they were a small group to begin with, and the game data was only available for 10 of them), they were excluded from the sample.

From the computer game survey, we utilized students' responses to the following questions: 'How often did you play the game?' (*1–2 times*, *3–5 times*, *6–10 times*, *11–20 times*, *more than 20 times*); 'Where did you play the game?' (*1 = at school*, *2 = at home*, *3 = at friends' home*, *4 = in the library*, *5 = somewhere else*, more than one alternative could be chosen); 'How did you find the I Know, I Can, and I Do components'? (*1 = very boring*, *2 = rather boring*, *3 = okay, it will do*, *4 = rather good*, *5 = excellent*); 'How much did the KiVa game help you to understand the matters presented in the KiVa student lessons?' (*1 = not all*, *2 = only little*, *3 = to some extent*, *4 = quite much*, *5 = very much*).

Results

Playing the game during the school year: how often, where, and with whom?

Most of the 1188 students in the final sample reported that they had played the game 3–5 times (43.1 per cent). The others had played 6–10 times (31.1 per cent), 1–2 times (10.3 per cent), 11–20 times (8.4 per cent) or more than 20 times (7.2 per cent). Chi square analyses indicated that boys had played significantly more than girls, fourth- and fifth-graders had played more than sixth-graders, and victims had played more than bullies and noninvolved children.

About half of the students, 52.2 per cent, had played the KiVa game only in school, 39 per cent had played both in and outside of school, and 8.8 per cent had played only outside the school. Girls, younger students (fourth- and fifth-graders) and victims were more likely to have played both in and out of school than boys, older students (sixth-graders) and nonvictimized students (bullies and non-involved children). The most typical places to play the game (besides school) were home (43.1 per cent of students), friends' homes (5.3 per cent), library (4.3 per cent) or some other place (5.3 per cent).

Overall, most students had played the computer game alone (84.5 per cent) but some had played together with their friends (26.3 per cent), siblings (6.2 per cent), parents (2.9 per cent) or with someone else (6.2 per cent). Girls had more often played alone than boys, fourth-graders had played with their siblings more than fifth- and sixth-graders, and, interestingly, bullies had played with their parents more than the other two groups.

How were the I Know, I Can, and I Do modules perceived by the students?

Overall, the students liked the anti-bullying computer game. The average values (on the scale from 1 to 5) given by them were 3.57 (*I Know*), 4.11 (*I Can*), and 3.49 (*I Do*) for the different modules of the game, respectively. Observing the mean scores suggests that *I Can*, which is the most game-like module where students move around in the virtual school, meet people and make behavioural choices, was liked the most.

A multivariate analysis of variance (MANOVA) was conducted where sex, grade level, and the child's bullying status (bully, victim, non-involved) were used as categorical independent variables and the perceptions of *I Know*, *I Can*, and *I Do* components of the game as the dependent variables. The main effects were tested first, and after that all possible two-way interactions (sex X grade, sex X bullying status, grade X bullying status) were added to the model.

According to the multivariate test of main effects, all independent variables (sex, grade, and bullying status) had a significant effect on students' perceptions of the game. The univariate tests further indicated that sex influenced especially the perceptions of the *I Do* module ($F(1) = 16.86, p < .001$), grade had an effect on perceptions of all three modules (*I Know* $F(2) = 19.504, p < .001$; *I Can* $F(2) = 32.36, p < .001$; *I Do* $F(2) = 21.54, p < .001$), and bullying status had an effect on how students perceived the modules *I Know* ($F(2) = 4.93, p < .01$) and *I Do* ($F(2) = 5.35, p < .01$). Girls liked the *I Do* module more than boys, fourth-graders liked each module more than fifth-graders, who in turn liked them more than sixth-graders. Finally, victims of bullying liked both the *I Know* and *I Do* modules more than non-victimized children. Typically, non-involved children tended to like the game relatively much and bullies tended to like it least. None of the two-way interaction effects was significant.

Finally, we created scores reflecting students' overall liking of the KiVa computer game by averaging across students' evaluations of the *I Know*, *I Can*, and *I Do* modules and ran another analysis of variance predicting variation in these scores. Grade level ($F(2) = 33.73$, $p < .001$) and bullying status ($F(2) = 5.74$, $p < .01$) turned out as significant predictors: fourth-graders gave the most favourable evaluation of the game, followed by fifth-graders, followed by sixth-graders (each group differing from each other significantly), and victims liked the game significantly more than the non-involved children who in turn liked it significantly more than bullies (Figure 14.2).

Did the computer game help the students understand the matters presented in KiVa student lessons?

On average, students felt that the KiVa game helped them to understand the matters presented in KiVa lessons (average score = 3.35 on the scale from 1 to 5, between response alternatives 'to some extent' and 'quite much'). There were significant main effects of sex ($F(1) = 38.99$) and grade ($F(2) = 12.51$) on the extent to which students perceived the game beneficial: girls perceived the game as more beneficial than boys, and fourth- and fifth-graders felt they benefited from it more than sixth-graders did. Bullying status was unrelated to whether students felt the game helped them to understand the issues presented during the student lessons. None of the two-way interaction effects added to the model was significant.

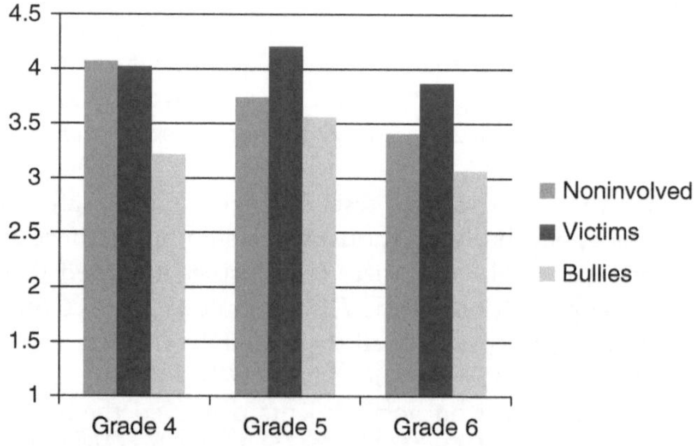

Figure 14.2 Overall perceptions of the KiVa computer game by victim, bullies, and non-involved children in grades 4–6

Conclusion

The KiVa anti-bullying computer game seemed to sit well as part of the more traditional remedies to tackle bullying. The program developers had recommended that students play during the five lessons during the school year but also (if possible) between the lessons outside the school. Half of the students had played the game only in school and 10 per cent only outside the school. The latter was probably due to the fact that some schools did not have a computer lab where everyone could play in the same time. As many as 40 per cent of the students reported having played the game both in and out of school. The high amount of voluntary playing outside the school suggests that students in general found the game positive, worthy of spending their leisure time on. We were impressed to discover that children who were victimized by their peers played the game more often both in and out of the school compared with bullies and non-involved students.

There was a pattern that emerged repeatedly in our findings: girls, younger students, and victims of bullying perceived the game more positively than boys, older students and non-victimized children did. In the final version of the KiVa program, the version of the game that was evaluated by students in the present study is indeed targeted at fourth-graders. Perhaps older students, especially sixth-graders, perceived it somewhat childish in comparison to other virtual games they are used to. When it comes to victimized students, it might be that they were relieved to discover that the game as well as the whole KiVa program is supportive of them. Hopefully they also gleaned some ideas of how to respond to bullying. In the KiVa program, the main focus is indeed on helping the victims: encouraging peers to support the victims instead of reinforcing the bully, and providing victims with efficient strategies to use in bullying situations. By using the computer game the victims could, in private, test themselves with respect to different characteristics (What are my best ways to resist bullying?), practise new skills in a virtual school (What to do if I get bullied?), and report which ones of the KiVa rules they had put into practice (I have reported bullying to adults). They also received feedback of their performance in a neutral way via the computer.

This year, final modifications of the Kiva computer games are being undertaken, based on the feedback we have received from teachers and students. We will add to the KiVa game a component meant especially for the victimized students: they will be offered information of where and how to seek help if victimized. There will also be a virtual mailbox through which the victims can report bullying directly to the school's KiVa team. Such a team exists in each school implementing the KiVa program, and their task is to tackle the cases of bullying coming to attention.

The present study was about students' use and perceptions of the game. With respect to future research, our data will enable us to study the effects of playing the game on students' *actual attitudes and behaviours*, by comparing changes taking place in children who played vs. who did not play the game (or children who played only a little vs. a lot) during the intervention year. That will shed light on whether the game truly enhanced the positive effects obtained with the KiVa program (reducing bullying and victimization, increasing constructive bystander behaviours) or whether it mainly contributed to motivating students and engaging them in other program components as well.

References

Anderson, C.A. and Bushman, B.J. (2001) 'Effects of violent games on aggressive behavior, aggressive affects, psychological arousal and prosocial behavior: a meta-analytic review of the scientific literature', *Psychological Science*, 12: 353–9.

Juvonen, J., Graham, S. and Schuster, B. (2003) 'Bullying among young adolescents: the strong, weak, and troubled', *Pediatrics*, 112: 1231–7.

Kärnä, A., Voeten, M., Little, T., Alanen, E., Poskiparta, E., and Salmivalli, C. (under review) 'Effectiveness of the KiVa anti-bullying program in grades 1–3 and 7–9'.

Kärnä, A., Voeten, M., Little, T., Poskiparta, E., Kaljonen, A., and Salmivalli, C. (2011) 'A large-scale evaluation of the KiVa anti-bullying program: grades 4–6', *Child Development*, 82: 311–30.

Olweus, D. (1994) 'Bullying at school: long-term outcomes for the victims and an effective school-based intervention program', in L.R. Huesmann (ed.) *Aggressive Behavior: Current Perspectives*, New York: Plenum Press, pp. 97–130.

——(1996) *The Revised Olweus Bully/Victim Questionnaire*, Bergen: Research Center for Health Promotion (HEMIL Center), University of Bergen.

Rodkin, P., Farmer, T., Pearl, R. and Van Acker, R. (2006) 'They're cool: social status and peer group supports for aggressive boys and girls', *Social Development*, 15: 175–204.

Salmivalli, C. (2010) 'Bullying and the peer group: a review', *Aggression and Violent Behavior*, 15: 112–20.

Salmivalli, C., Kärnä, A. and Poskiparta, E. (2010) 'From peer putdowns to peer support: a theoretical model and how it translated into a national anti-bullying program', in S. Jimerson, S. Swearer and D. Espelage (eds) *The Handbook of Bullying in Schools: An International Perspective*, New York: Routledge, pp. 441–54.

Salmivalli, C., Lagerspetz, K., Björkqvist, K., Österman, K. and Kaukiainen, A. (1996) 'Bullying as a group process: participant roles and their relations to social status within the group', *Aggressive Behavior*, 22: 1–15.

Salmivalli, C., Voeten, M. and Poskiparta, E. (2011) 'Bystanders influence the frequency of bullying in a classroom'. *Journal of Clinical and Consulting Psychology*, 40: 668–76.

Sapouna, M., Wolke, D., Vannini, N., Watson, S., Woods, S., Schneider, W. *et al.* (2010) 'Virtual learning intervention to reduce bullying victimization in primary school: a control trial', *Journal of Child Psychology and Psychiatry*, 51: 104–12.

Sherry, J.L. (2001) 'The effects of violent video games on aggression: a meta-analysis', *Human Communication Research*, 27: 409–31.

Sijtsema, J., Veenstra, R., Lindenberg, S., and Salmivalli, C. (2009) 'An empirical test of bullies' status goals: assessing direct goals, aggression, and prestige', *Aggressive Behavior*, 35: 57–67.

Smith, P.K., Morita, J., Junger-Tas, D., Olweus, D., Catalano, R. and Slee, P. (1999) *The Nature of School Bullying: A Cross-National Perspective*. London: Routledge.

Solberg, M. and Olweus, D. (2003) 'Prevalence estimation of school bullying with the Olweus bully/victim questionnaire', *Aggressive Behavior*, 29: 239–68.

Chapter 15

FearNot!

An innovative interdisciplinary virtual intervention to reduce bullying and victimization

Dieter Wolke and Maria Sapouna

Positive uses of information technology

Popular media such as computers and mobile phones are tools that provide a range of opportunities for teaching students how to cope with bullying and aggression in a more engaging way than traditional curriculum-based approaches (Padmore *et al.*, 2006). A recent international systematic review of school-based programmes aimed at reducing bullying and victimization found that half of the evaluated interventions incorporated computer-based learning tools and these programmes showed the biggest reductions in bullying victimization (Farrington and Ttofi, 2009). In the majority of cases, the computer-based element of the programme comprised non-interactional educational videos showing bullying situations or videotaped interviews with victims of bullying (Baldry and Farrington, 2004; Raskauskas, 2007).

A more recent development in the use of media technologies as a therapeutic tool is virtual learning interventions that allow users to learn and practise problem-solving and coping skills by interacting in a safe virtual environment that simulates real life situations in a highly believable way (for a recent example, see Petrasova *et al.*, 2010). This type of intervention has recently started to be used to treat individuals experiencing a range of psychological disorders such as phobias, eating disorders, substance abuse and post-traumatic stress disorder with advantages over more conventional interventions (Riva, 2005). The basic feature of these programs is a computer-based environment that creates a highly believable learning experience for users who find themselves 'present' in the situation that causes emotional distress and, as a result, learn experientially how to deal with it in a shorter period of time (ibid.).

The FearNot! intervention

The intervention presented in this chapter provides an innovative example of the use of virtual learning technologies to teach children how to cope with bullying effectively. The developed computer game, called FearNot! (Fun with Empathic Agents to Reach Novel Outcomes in Teaching), possesses a unique set of properties that distinguishes it from other computer-based interventions in the field of bullying. First, the computer characters in FearNot! are intelligent agents that are able to interpret and feel emotions, and indeed act in a way that reflects their internal emotional state (Louchart *et al.*, 2006). In this way, characters are more believable and children are more able to empathise and engage with them. Furthermore, FearNot! creates a highly personalized and interactive experience for the user as the story is not scripted and predetermined but evolves every single time according to the advice given by each child, therefore increasing the believability of the game. These unique characteristics of FearNot! make it easier for children to learn experientially and transfer the knowledge they have acquired in the virtual school to real life contexts. A detailed description of the technological aspects of the English and German version of FearNot! is available elsewhere (Aylett *et al.*, 2006; www.e-circus.org.uk).

FearNot! comprised the sole component of an anti-bullying intervention that was targeted at primary school students between the ages of 8 and 11. When playing this game, children watch a number of bullying incidents that occur between animated characters in a virtual school and at the end of each episode they are asked to interact with the victim character who is asking them for advice on what to do to avert similar situations in the future (Figure 15.1). Children can 'talk' to the animated students of the virtual school by typing in text that is understood by the virtual characters who subsequently respond to the advice given according to their emotional state and how powerful they feel to follow the child's advice, as would happen in real life. When the animated victim character decides to accept the advice given, an episode follows that shows the outcome of the action taken.

Figure 15.1 FearNot! screenshots of male verbal bullying episode

Outcomes were designed to reflect real-world probabilities of success of the most commonly used coping strategies against bullying as evidenced in the literature, so, for example, crying as a response to bullying would never appear to be successful whereas making a new friend is almost always successful. In this way, children are able to learn effective strategies of dealing with bullying victimization in different situations.

The remainder of this chapter describes the results from the evaluation of FearNot! that was conducted in the UK and Germany. The aim of the evaluation was to ascertain whether use of FearNot!: (1) helped victims of bullying escape victimization; (2) reduced rates of victimization in intervention classes; and (3) increased the density and reciprocity of ties among students in intervention classes.

The FearNot! Evaluation

In 2007, 39 primary schools in the UK (Warwick, Coventry and Hertfordshire) and Germany (Bavaria, Hesse) were invited to participate in the FearNot! evaluation. Twenty-seven schools agreed to participate in the project, resulting in an eligible sample size of 1178 students. Baseline data were collected in November 2007. Schools with computers able to run the FearNot! software well were assigned to the intervention group (n = 13, total number of eligible students = 555) and schools whose computer systems did not meet the technical requirements of FearNot! were allocated to the control group (n = 15, total number of eligible students = 623). The first post-test was done one week after the intervention finished and the second post-test was completed four weeks after the intervention. Of the 1178 students who were eligible to participate in the study, 49 were not given consent to take part, rendering an initial sample of 1129 students. As can be seen in Figure 15.2, the total number of students who took part in all assessments and whose responses were, therefore, analyzed as part of this study were 942 (83.4 per cent of the eligible sample of 1,129), 455 in the intervention group and 487 in the control group. Among these, 494 (52.4 per cent) were male and 448 (47.6 per cent) were female. UK pupils (n = 514) were older than German pupils (n = 422) (9.36 vs. 8.34; $t = 28.71$; $p < .001$, missing data for 6 pupils).

Measurements

All participants in the intervention and control groups were given a brief awareness-raising session on the type of behaviours that are defined as bullying before they were asked to fill in the baseline questionnaires. The same questionnaires were filled in at the first and second post-test, one and four weeks after the intervention respectively. Questionnaires took no longer than an hour to complete and were available in English and German.

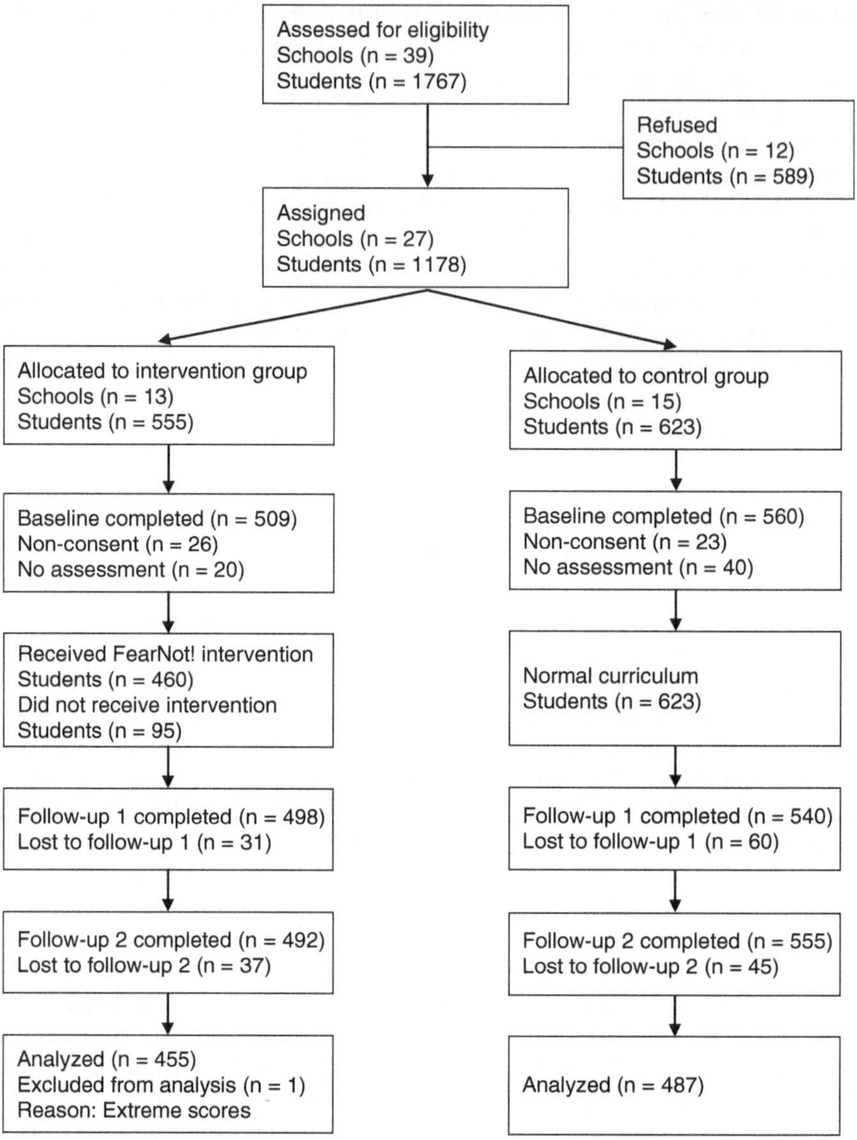

Figure 15.2 Participant flow diagram
Note: In one school, one class was able to take part in the intervention group while the other was included in the control group.

Data was collected on participants' gender, age, number of siblings and living arrangements (i.e. with whom they lived at home). The main individual-level outcome was victimization (experienced and escaped). Children were asked to indicate how often they had experienced either direct or relational victimization over the last month. Direct victimization included being hit/beaten up, having things stolen, being threatened/blackmailed, being called nasty names and having nasty tricks played on them. Relational victimization included being left out of games, having children telling them they don't want to be their friend any more and having nasty lies or rumours spread about them. Children were classified as victims if they had experienced direct and/or relational bullying 'more than 4 times' or 'at least once a week' in the last month. Children were categorized as escaped victims if they reported being directly and/or relationally bullied at baseline, but had experienced neither form of bullying at the first (T1 escaped victim) or the second follow-up (T2 escaped victim).

We also wanted to assess if the intervention would help improve the class climate as indicated by an increase in the number of one-way and reciprocal relationships formed among students. To test this assumption, we asked children to provide up to three nominations of their best friends. Based on these nominations, we were then able to derive a measure of class density and reciprocity using the Ucinet social network software (Borgatti *et al.*, 2004). *Density* was measured as the number of ties present in the class divided by the number of all possible ties. *Reciprocity* was the proportion of reciprocated ties in the network (relative to the number of actual ties). For more details of the methodology of the study, see Sapouna *et al.* (2010).

Main findings

As we were not able to randomize, it was important to check that there were no significant baseline differences between the intervention and the control group in terms of measured characteristics. Our analyses indicated that participants in the intervention and control groups were comparable at baseline in terms of gender, age, living arrangements and family composition. In addition, intervention and control classes were similar with respect to size and socio-economic status. There were fewer victims among the intervention group children compared with controls at baseline but the difference was not statistically significant. Both groups also reported similar rates of bullying perpetration. Comparisons between the two countries showed that UK and German children experienced similar rates of victimization (28.6 per cent vs. 23.6 per cent; $\chi^2_1 = 2.82; p = .09$). However, there was a significantly higher number of bullies among UK children than German children (99; 20.3 per cent vs. 15; 3.7 per cent; $\chi^2_1 = 54.47; p < .001$).

As only 83.4 per cent of the initial eligible sample completed all assessments and were, therefore, analysed for the purposes of this study, we investigated whether non-responders differed in some significant way from responders. The children who missed follow-up 1 were similar to completers in terms of gender, age, living arrangements, family composition and baseline victimization. However, the children who were lost at follow-up 2 were significantly less likely than responders to be living with their mothers (90.6

per cent vs. 98 per cent; $\chi^2_1 = 13.68$; $p < .001$) and younger (8.68 vs. 8.91; $t = 2.41$; $p = .02$) at baseline.

After having ascertained that our intervention students did not differ significantly from control group students, especially in terms of levels of victimization experienced, we went on to test whether the software helped those students who were victimized at baseline to escape victimization at subsequent assessment points. What we found was that baseline victims in the intervention group were significantly more likely, compared to baseline victims in the control group, to be classified as escaped victims (i.e. non-victim at follow-up 1 but victim at baseline) one week after the intervention was completed (adjusted RR = 1.41, 95% CI: 1.02–1.81; n = 230). More specifically, out of 106 baseline victims in the intervention group, 53 (50 per cent) managed to escape victimization at follow-up 1, compared to 44 out of 124 (35.5 per cent) baseline victims in the control group. To provide some support for the cause–effect relationship of the intervention to this outcome, we investigated whether children who interacted longer or more actively with the software achieved better outcomes. Indeed, it was found that children who took part in a greater number of interaction episodes with the FearNot! characters were more likely to escape victimization at follow-up 1 (adjusted OR, 1.09; 95% CI, 1.003–1.18). However, a dose–response relationship was not found for total time of interaction with the software and total number of episodes watched. This means that what really mattered was the amount of active interaction children had with the characters and not simply how much time they spent playing with the software or watching episodes. Although results of the FearNot! intervention were positive when victimization status was re-assessed one week after the intervention was completed, a significant treatment effect was not maintained four weeks after the intervention.

The second question we examined was whether the intervention reduced overall rates of victimization in classrooms. In relation to this question, we found that the software worked better with UK students than German students. That is, UK children in the intervention group experienced significantly lower victimization rates at the first-follow up compared with controls (adjusted RR = 0.60, 95% CI: 0.36–0.93; n = 470). A similar trend was found in the overall sample but it was not statistically significant. However, replicating findings on escaped victimization, treatment effects were not maintained at the second follow-up.

Finally, we tested whether the intervention had the potential to have a positive effect on the number and quality of relationships formed in the classroom as children became more aware of bullying issues and more sensitized to the plight of vulnerable victims. To test this hypothesis, we assessed whether class density and reciprocity scores changed over time differently for intervention and control group classes. At baseline, intervention classes had a slightly higher mean density score compared to control group classes (0.11

vs. 0.9; $p < .05$). Similarly, the mean reciprocity score was higher in intervention classes compared to control group classes; however the difference was not statistically significant (0.29 vs. 0.24; ns). On average, 29 per cent of ties in intervention classes were reciprocal compared to 24 per cent of ties in control group classes. A two-way repeated measures ANOVA showed that density scores decreased over time for both groups therefore the intervention did not seem to have an effect. Similarly, the two-way repeated measures ANOVA for reciprocity demonstrated no significant main or interaction effects.

Implications

This chapter reported the results of the evaluation of a new virtual learning environment for exploring issues related to school bullying called FearNot!. The aim of this short-term intervention was to assess the benefits of using virtual learning computer programmes that simulate real life bullying incidents as a forum of experiential learning about bullying. We hypothesized that FearNot! would: (1) help victims of bullying escape victimization; and (2) improve classroom climate by reducing overall levels of victimization and promoting the formation of dense and reciprocal ties among students. These latter outcomes were included in the scope of the evaluation as it was assumed that interaction with FearNot! would promote children's general knowledge about bullying and enhance their empathy towards victims. By acting as a 'buddy' to the virtual victim, children were given the opportunity to feel the plight of victimized children and understand their position.

FearNot! was used by intervention students once a week for three consecutive weeks and despite the short duration of the intervention, results were positive. The children who benefited the most from the intervention were those that were self-identified as victims at the start of the study. Out of 106 baseline victims in the intervention group, 53 (50 per cent) did not report victimization one week after the intervention, compared to 44 out of 124 (35.5 per cent) baseline victims in the control group. This effect, however, was not maintained three weeks later.

These results are in line with the program's theory of change. FearNot! was primarily targeted for victims of bullying and it aimed to bring about change by raising victimized children's self-efficacy to deal with bullying. This was thought to be achieved by allowing children to vicariously experience highly believable bullying situations in a safe environment and discover more and less effective ways of responding to victimization. While playing FearNot! children acted as a friend to a virtual victim who needed advice on what to do to avert further bullying from peers, and, in doing so, were able to enhance their repertoire of coping strategies and to see first-hand which of the strategies they suggested were the most and least successful. It was assumed that, in the short-term, such knowledge would bolster children's confidence in their ability to respond effectively to bullying and, in the long-term, children would feel confident enough to transfer this experiential learning to real-life contexts, implementing successful coping strategies against bullying that would help them escape victimization.

FearNot! proved less successful in changing overall levels of victimization in classrooms and had no effect on improving the cohesiveness of children's networks. One reason for this might be that FearNot! was primarily designed for use with children experiencing bullying. Further, we postulate that the intervention would be in a better position to positively affect classroom climate, through reducing the risk of victimization and increasing the density and reciprocity of ties among students, if it included more than one component, targeted a wider audience (e.g. bullies, parents and teachers) and lasted longer.

Another important factor that may explain the limited success of the intervention, in particular in Germany, is reduced implementation fidelity and commitment from teachers. While in the UK the software was well accepted by teachers, in Germany the teachers more often expressed negative attitudes towards the software rating it more often as very poor or poor or saying that they would definitely or probably not consider using it again in their classes. German teachers were much less experienced in using computers in the classroom compared to UK teachers and faced more difficulties responding to technical problems. They were less well trained in IT and had a lack of IT support provisions in their schools. Therefore, the use of virtual learning may be limited to learning environments where IT use is part of the school learning experience with appropriately trained teachers. Interventions are unlikely to work unless they can be successfully implemented without placing too much burden on those responsible for delivery and sufficient motivational work takes places at the start of the programme to ensure buy-in from all relevant stakeholder groups (e.g. teachers, head principals, parents; Rohrbach *et al*, 1996).

On the whole, the results from the evaluation of the FearNot! anti-bullying intervention provide support for the use of computer-based programmes to tackle bullying at schools. The positive effects of incorporating computer-based learning tools in anti-bullying interventions have been highlighted in a recent systematic review of anti-bullying programmes that found the biggest reductions in bullying victimization to have been produced by interventions with a computer-based learning element (Farrington and Ttofi, 2009). These findings are encouraging in light of the growing use of computers in education and the development of new anti-bullying interventions that rely on virtual learning environments such as the KiVa program in Finland (Salmivalli *et al.*, 2009) (see Chapter 14).

Acknowledgements

This work was partially supported by the European Community and was partly funded by the eCIRCUS project IST-4-027656-STP with university partners Heriot-Watt, Warwick, Hertfordshire, Sunderland, Bamberg, Augsburg, Wuerzburg plus INESC-ID and Interagens. Thank you to the whole eCircus team (see http://www.macs.hw.ac.uk/EcircusWeb/) who contributed to the development of FearNot! and its evaluation. The authors are solely responsible for the content of this publication. It does not represent the opinion of the European Community, and the European Community is not responsible for any use that might be made of data appearing therein.

References

Aylett, R.S., Louchart, S., Dias, J., Paiva, A., Vala, M., Woods, S. and Hall, L. (2006) 'Unscripted narrative for affectively driven characters', *IEEE Journal of Graphics and Animation*, 26: 42–52.

Baldry, A.C. and Farrington, D.P. (2004) 'Evaluation of an intervention program for the reduction of bullying and victimization in schools', *Aggressive Behavior*, 30: 1–15.

Borgatti, S.P., Everett, M.G. and Freeman, L.C. (2004) *UCINET 6 for Windows: Software for Social Network Analysis*. Cambridge, MA: Harvard Analytic Technologies.

Farrington, D.P. and Ttofi, M.M. (2009) 'School-based programs to reduce bullying and victimization', *Campbell Systematic Reviews*, 2009: 6.

Louchart, S., Aylett, R., Enz, S. and Dias, J. (2006) 'Understanding emotions in dramas: a step towards interactive narratives', *Proceedings of Adaptation in Artificial and Biological Systems, 3, AISB 2006 Conference*, 38–44.

Padmore, M., Hall, L., Hogg, B. and Paley, G. (2006) 'Reviewing the potential of virtual learning environments in schools', in R. Aylett, T. Diener and Z. Pan (eds) *Edutainment*, Amsterdam: Springer-Verlag.

Petrasova, A., Czanner, S., Chalmers, A., Farrer, J. and Wolke, D. (2010) 'The playability evaluation of virtual baby feeding application', *Proceedings of 2010 Second International Conference on Games and Virtual Worlds for Serious Applications*, pp. 95–100.

Raskauskas, J. (2007) *Evaluation of the Kia Kaha Anti-Bullying Programme for Students in Years 5–8*, Wellington: New Zealand Police.

Riva, G. (2005) 'Virtual reality in psychotherapy: review', *Cyber Psychology and Behaviour*, 8(3): 220–40.

Rohrbach, L., D'Onofrio, C., Backer, T. and Montgomery, S. (1996) 'Diffusion of school-based substance abuse programs', *American Behavioral Scientist*, 39: 19–34.

Salmivalli, C., Kärnä, A. and Poskiparta, E. (2009) 'From peer putdowns to peer support: a theoretical model and how it translated into a national anti-bullying program', in S. Shimerson, S. Swearer and D. Espelage (eds) *The Handbook of School Bullying: An International Perspective*, Mahwah, NJ: Lawrence Erlbaum.

Sapouna, M., Wolke, D., Vannini, N., Watson, S., Woods, S., Schneider, W., *et al.* (2010) 'Virtual learning intervention to reduce bullying victimization in primary school: a controlled trial', *Journal of Child Psychology and Psychiatry*, 51(1): 104–12.

Chapter 16

Using mobile phones to counter cyberbullying

An innovative project

Alison J. Wotherspoon, Greg Cox and Phillip T. Slee

Introduction

This chapter describes a project called the *Anti-Cyberbullying Phone Project* that was funded through the 2006 Australian National Safe Schools Week, and supported by *The Coalition to Decrease Bullying, Harassment and Violence in South Australian Schools* in South Australia. The project brought together young people from 16 South Australian high schools for a series of workshops, where they created a number of positive, animated, anti-bullying messages for distribution via mobile phones. This innovative project taught young people a range of new skills and engaged them in developing resources suitable for use in bullying and cyberbullying prevention and intervention programmes.

The workshop was made possible by significant initiatives undertaken by the Australian Commonwealth and State governments over the past decade. To provide a background to this innovative project a brief overview of the national context is useful. In July 2003 the Australian Federal Ministerial Council for Education, Early Childhood and Youth Affairs (MCEECDYA) endorsed the National Safe Schools Framework (NSSF), paving the way for Australian education authorities to develop a consistent national approach to bullying, violence and child protection in all Australian schools to ensure the physical and emotional safety of students. The National Safe Schools Framework is a set of nationally agreed principles for safe and supportive school environments, that recognizes the importance of a whole school approach to bullying, violence and child protection, and the need to develop and implement policies and programs in consultation with the wider school community.

The NSSF was delivered to schools in Australia in 2004, and in July 2005 the Government of South Australia announced the creation of *The Coalition to Decrease Bullying, Harassment and Violence in South Australian Schools* (the Coalition) whose role was to provide advice to the schooling sector in general and the Minister for Education and Children Services. This organization brought together, for the first time, the three schooling sectors in the state:

the South Australian Department of Education and Children's Services (DECS) responsible for the state/government school system, Catholic Education South Australia (CE-SA) and the Association of Independent Schools of South Australia (AISSA), and allowed them to work innovatively and collaboratively together to implement the NSSF and other initiatives.

Four South Australian academics, specializing in the field of bullying and violence, were invited to join the Coalition to advise the schooling sector of current research and to identify areas of interest for future research: Professors Phillip Slee (Flinders University) and Ken Rigby (University of South Australia) and Drs Barbara Spears (University of South Australia) and Shoko Yoneyama (University of Adelaide). At a later date a representative from the South Australian Police was also invited to join the Coalition, further extending its reach into the community. Based on this South Australian model and with advice from Professor Rigby, the Queensland Coalition of Schools recently established the Queensland Schools Alliance Against Violence (QSAAV) in February 2010.

To mark the beginning of the NSSF, the Commonwealth Government funded National Safe School Weeks in 2006 and 2007 and the first *Anti-Cyberbullying Phone Project* was undertaken during 2006. A similar workshop ran in Adelaide during the 2007 National Safe School Week and built on the success of the first workshop. Only the first workshop is discussed in this chapter.

The Australian context and background to the anti-cyberbullying phone project

Cyberbullying is a significant problem for young Australians as they live in a nation of sophisticated ICT users (Campbell and Gardner, 2005). The majority of Australians, with home Internet access, use broadband technology, and in December 2007 there were 7.1 million Internet subscribers, of which 6.14 million were household subscribers. In 2007, 64 per cent of Australian households had home Internet access, and 73 per cent of the population had access to a home computer. Of the 2.7 million Australian children aged between 5 and 14, in 2007, 65 per cent were using the Internet and 92 per cent were using a computer (Australian Government, Department of Foreign Affairs and Trade, 2010). The national household broadband access had reached 62 per cent in 2008–9 (http://www.abs.gov.au/ausstats/abs@.nsf/Products/4901.0~Apr+2009~Main+Features~Internet+use+and+mobile+phones?).

Cyberbullying is not so much a unique, and new form of bullying, but rather it is a form of bullying that uses new tools and technology. Bullying can, and does, impact on some young people 24/7, via the internet, through email, social networking and with the use of mobile phones (Beran and

Li, 2004, 2005). The Anti-Cyber Bullying Phone Project was designed to empower young people to use their mobile phones creatively and allow them to use mobile phone technology in an imaginative and constructive way. The mobile phone technology that was available when this project took place in 2006, was nowhere near as powerful, as versatile or as prevalent as what is available today. However, it is the innovation at the time, and the positive use to which this technology was put, that has made a significant contribution to our understanding of how young people can be involved in their own learning about the creation of anti-cyberbullying messages.

In 2008, Australia ranked fourth out of 70 countries for its 'e-readiness' and was ranked first overall in the category of social and cultural environment including aspects such as level of education and literacy, level of internet literacy, degree of entrepreneurship, technical skills of workforce, and degree of innovation (Australian Government, Department of Foreign Affairs and Trade, 2010). E-readiness is a measure of a country's information and communications technology (ICT) and the ability of businesses and consumers to use ICT to their advantage.

Mobile phones in Australia

Australia has one of the highest rates of mobile phone ownership in the world (Australian Government, Department of Foreign Affairs and Trade, 2010). Since the 2006 workshop, mobile phone technology has changed significantly with the advent of Smart phones. Smart phones are effectively hand-held computers integrated with mobile phone functions. In 2009, data was collected on children's mobile phone ownership in Australia and it was estimated that 31 per cent of children had a mobile phone. More girls than boys had a mobile (33 per cent compared to 29 per cent) and within the 12 to 14 year old age group 76 per cent of them had mobiles (http://www.abs.gov.au/ausstats/abs@.nsf/Products/4901.0~Apr+2009~Main+Features~Internet+use+and+mobile+phones). The majority of children said that the primary use of their phones was to keep in contact with their families. This primary use decreases as young people go into their teens where they use mobiles primarily to communicate with their friends.

Mobile phones can be therefore seen to be incredibly popular devices with young people in Australia. Given that they are powerful communication devices, they can also be used as very effective weapons when used by young people to bully and harass others. Bullying behaviour on mobile phones may include spamming another young person's phone with hundreds of constant unwanted phone calls and texts, and the rapid, viral dissemination of images, videos and comments (Campbell and Gardner, 2005; Beran and Li, 2004). This form of cyberbullying can, in extreme cases, result in a young person who is being targeted taking their own life. A notable recent example was a young college student in America committing suicide after being sexually outed via the internet and mobile phones (http://www.smh.com.au/technology/technology-news/sex-tape-suicide-had-he-been-in-bed-with-a-woman-this-would-not-have-happened-20101001-15zld.html).

Despite the negative ways in which mobile phones can be used, there are also opportunities to do creative work on them and disseminate this work widely. In theory, it is now easier than ever before for young people to use mobile phones creatively. Using an ever-increasing number of phone apps, young people (and old) can distribute almost any message, and any quantity of messages they wish, to a huge number of people affordably, effortlessly and immediately.

With the scale of distribution now available on mobile phones, it is timely to ask a number of questions about the creativity and technology in the current wave of enthusiasm for smart phones. First, where does creativity really fit? And, second, how do you teach people to take responsibility for material in cyberspace that has the potential to go viral overnight with potentially devastating results? Young people who have access to smart phones can easily create music, videos, and animations and distribute them immediately via their phones to their friends on social networking sites such as Facebook and MySpace.

There are interesting questions to explore about creativity and the use of this technology. If young people are simply using applications supplied or sold to them by mobile phone corporations, such as using an application to manipulate a picture of your friend's face to make them look fat, is this particularly creative? If you express yourself with a ringtone that is ready-made, does that make you unique? Is it artistic to make a statement out of pre-existing options available to purchase and download? And does this plethora of consumer options give the user a sense of ownership or make them a content developer or an artist? Possibly. Mixing together these elements could be considered creative practice but the real challenge may be whether it is possible to take control of the applications, experiment and manipulate them in ways that they may not have been intended to be used to create truly innovative and unique results. The Anti-Cyber Bullying Phone Project provided a model of how to be creative outside of predetermined applications and within a workshop environment that encouraged the participants to consider how the technology could be used responsibly and in positive ways.

The Anti-Cyber Bullying Phone Project

The Anti-Cyber Bullying Phone Project aimed to counteract cyberbullying by giving young people the skills to make their own creative content for mobile phones, that would tell their stories, present solutions to bullying and could be disseminated via bluetooth technology, freely, from mobile phone to mobile phone. The project was inspired by a workshop run by The

Phone-Book Limited (http://www.fonebk.com/ltdx/index.htm), a group of British creative artists, interested in teaching people how innovative content for mobile phones could be created cheaply, easily and distributed freely.

The Anti-Cyber Bullying Phone Project was a series of workshops designed to teach high school students to use mobile phone technology in positive ways.

Methodology

Letters of invitation were sent out by the Department of Education and Children's Services to a number of State, Catholic and Independent high schools, in South Australia, inviting them to participate in the Anti-Cyber Bullying Phone Project. The offer to participate was taken up by 16 South Australian high schools, and 72 students and 15 teachers attended. The participants attended one of four one-day workshops run between the 15th and 19th May 2006 at the Technology School of the Future in Hindmarsh, Adelaide. They were subsequently invited to present their work to the other participants at a presentation day on the 16th June 2006 at Flinders University.

The workshop process

Before the workshop, letters were sent out to each participating school outlining the structure and intent of the workshops. Each school was responsible for choosing the students and teachers who were to attend.

Since the workshop focused on bullying, the schools were encouraged to work with the students attending, on bullying issues in advance of the workshop. There was an expectation by the workshop coordinators that the ideas and story concepts the students wanted to tell would have been discussed and identified at the school and in advance of the workshop. This did not happen in a number of the participating schools.

Approximately four students from each school with one or two teachers attended the workshops at the Technology School of the Future (TSOF), a government-funded school resource, which provided technology suites and recording studios. The workshops took place in computer rooms where the students had access to computers and space to work in groups. The technical training was devised and delivered by a digital artist, Sasha Grbich, who was employed by ANAT (Australian Network for Art and Technology). Sasha's background and involvement in the project can be viewed at: http://www.bullyingnoway.com.au/talkout/profiles/practitioners/sashagrbich.shtml.

The workshop was divided into four parts. In the first session the students were introduced to the technology and the way in which they could make and distribute their work via mobile phones or by uploading work onto the internet. In this introductory session there was also a discussion of how mobile phone content could be used in a positive way to combat bullying and as a response to cyberbullying on mobile technology. The second part focused on the practical skills required to make a short stop-motion animation. The students were introduced to the concept of devising a stop frame animation with an anti-bullying message, told how to storyboard their ideas, and were then given inexpensive materials to work with, ranging from plasticine to paper and coloured pens, and taught how to capture their shots with a camera to animate them. The third part gave the students time to discuss and develop their concept, and to devise and make a short animation or video that showed a creative and positive way to deal with bullying. The students were then able to take their enhanced knowledge of media production and ICT (Information and Communication Technologies) back to their schools where it was envisaged that they would be able to teach their teachers and peers new skills.

Given the limited time of the workshop, each school was given a pack that included all software used, notes from the day and a bluetooth dongle so they could complete the post-production at a later date at their school and be able to convert the files to formats that could be distributed through mobile phones or uploaded onto the internet.

The creative works

The works produced by the students varied greatly in form, execution and content. They ranged from simple stories about being the target of a bully due to being different, to quite sophisticated issues concerned with alcohol, date rape and the distribution of inappropriate images via a mobile phone. Not all of the animations were successful and not all of them were completed due largely to a range of technical problems. Seven, however, were eventually completed and put up on the Bullying. No Way! website and four are briefly described below.

Hilda was created by a team of students at St George College, South Australia and told the story of Hilda, a hippopotamus who is the target of bullying (http://www.bullyingnoway.com.au/talkout/spotlight/vid_hilda.shtml). Although silent, it offers an insight into the torment and exclusion students feel when bullied about their physical appearance.

Bully was made by students from Banksia Park International High School, South Australia, who used plasticine characters and detailed backdrops, to produce a creative mobile movie that targets bullying within the classroom in society and 'Say NO to Bullying!' (http://www.bullyingnoway.com.au/talkout/spotlight/vid_bully.shtml).

Sushi Pie, created by a team of students from Cabra Dominican College, South Australia, is an ingenious mobile movie which directly targets cultural differences and shows how compromise can have a much more interesting outcome than conflict (http://www.bullyingnoway.com.au/talkout/spotlight/vid_sushi.shtml).

Caught on Film targets bullying while also addressing teen drinking and how vulnerable people are if they are in an intoxicated state. The clip was created by a team of students at University Senior College, South Australia and shows the need for young people to act responsibly (http://www.bullyingnoway.com.au/talkout/spotlight/vid_caught.shtml).

All of the completed works are currently on-line and downloadable from the Australian-based national Bullying. No Way! website (www.bullyingnoway.com.au/) at http://www.bullyingnoway.com.au/talkout/spotlight/mobilemovies.shtml.

What emerged strongly from the workshops was the way in which the young people were comfortable with the technology and that many of the teachers were not.

The challenges

The workshops were extremely innovative at the time but were not without their challenges. Within the workshops the digital artist and students were trying to do things that now, can be effortlessly achieved, but in 2006 were technically challenging and a fairly radical use of the emerging technologies. There were problems with software programs, compatibility of computers and encoding issues. In 2006, creating content for mobile phones was a relatively new area of creative practice and there were no industry standards across mobile and wireless devices. As much as possible, the workshop aimed to use freeware or shareware programs, so that all schools could have access to the technology. Free trial software programs were used to encode the final edited film and animation files to a 3gp format for mobile phones, but it was found during the workshops and in the follow-up session in schools that a number of the programs did not consistently do what they were meant to do and were not stable.

The choice of which students and teachers attended impacted on the outcomes. With the workshop focus on bullying, many of the teachers sent by the schools were counsellors or staff dealing with bullying issues rather than teachers interested in or comfortable with technology either in the form of computers or mobile phones.

There was a great disparity between the schools involved in regard to how media was taught and supported, as well as extremely varied access to both software and hardware. After the initial workshop at TSOF some schools committed clearly defined class times to finish up the work while others had to try to complete the project during lunch breaks and after school. Editing and sound post production software and hardware available to the students varied greatly from school to school and the students who had been accompanied by counsellors rather than IT teachers were at a disadvantage. A longer lead time and clearer guidelines would have allowed schools to plan to send a mix of teaching staff that suited the way bullying interventions as well as ICT skill acquisition was addressed and delivered in the school. Although going on an excursion to attend a workshop at TSOF was fun and allowed the students and teachers to focus on the day without distractions, a similar program delivered within the school would have, in most cases, been more effective. The trainers would have been able to individualize the workshops to match the resources, both technical and human, available within each school and to deliver training that could have been more easily transferred, and repeated by staff within a school and would therefore be sustainable.

The follow-up

Ten of the sixteen schools presented their work at Flinders University four weeks after the workshops took place. The students were able to show their work to each other, discuss the making of it, and receive feedback on their work from their peers and the workshop coordinators.

Feedback

Feedback forms had been distributed to all schools and completed forms were received from seven teachers and twelve students. It was, in the main, very positive. Many of the students found that the workshop had made them think about different ways in which bullying could be addressed. Some students said that the workshops resulted in discussions taking place in their schools that went into more depth than was possible in the workshops. Most of the students felt confident with the programs taught and had enjoyed making the animations. A number of students wished that the workshops had gone over two days, as there was too much to cover in one day. The students found the most engaging aspects to be the making of the animations and seeing the other students' work at the presentation day (Grbich and Wotherspoon, 2006).

The teachers were very positive about the workshops and felt the project had given the students an opportunity to think about bullying in a relaxed atmosphere. Most felt comfortable with learning new technical skills but also believed they needed more technical support when they took their new knowledge back to their schools. Some noted that it would have been useful to include IT teachers in the workshops as the majority who attended were counsellors involved in bullying rather than those teaching technology.

An important part of the workshop ethos was the notion that ICT allows knowledge to be shared democratically. The dissemination of skills and information was delivered to those who participated in the workshops in a way in which they could in turn easily and affordably pass this knowledge on to others. It was a grassroots and community approach to the use of ICT and the transfer of knowledge. The access to, and use of, mobile phone technology is something that can be used in the poorest communities and can result in extraordinary outcomes, from political activism to highlighting individual civil rights. Many examples can be found at http://www.mobileactive.org/about.

During the presentation day a number of the school groups announced their intentions to repeat the animation workshop with groups of younger students in their schools and share the work they had made with the whole school with presentations at school assemblies.

The level of enthusiasm and interest shown by the students was high, as was the creativity shown in their work. The program was deemed by many of the participants to be a success and went on to be repeated, in a new and improved form which included workshops held in schools, the following year. The main criticism was that outcomes would have been better had there been more lead-up time, as the schools could have integrated the workshop more effectively into the whole year's curriculum.

The outcomes

The students presented their movies at several Australian conferences, notably the Safer South Australian Schools: Meeting the Challenges of Bullying Conference UNISA, Adelaide and the National Safe Schools Week, Australian Government Quality Teaching Practice Teacher Education Event, Rydges, Melbourne, which were both held in 2006. As well as the films being available online and downloadable on the Bullying. No Way! website http : www.bullyingnoway.com.au/ there is also a step-by-step guide for schools on how to run a mobile movie workshop (http://www.bullyingnoway.com.au/talkout/spotlight/makeyourown.shtml).

Conclusion

Bullying is a serious and ongoing challenge for schools to manage and for young people and families to deal with. Current research in Australia, as well as in Europe and North America, suggests that rather than increasing, there has been a reduction in school bullying over the last decade (Rigby, 2010). Programs and projects that help to inform young people, challenge their notions of what is acceptable and normal, empower them with a greater knowledge of the issues to do with bullying and harassment and offer them a range of new skills to employ are important in helping to change bullying cultures in schools. Innovative initiatives such as the Anti-Cyberbullying Phone Project can help make a difference.

References

Australian Government, Department of Foreign Affairs and Trade (2010) '*About Australia: Information and Communications Technology*', available at: http://www.dfat.gov.au/facts/ict.html (accessed 14 December).

Beran, T. and Li, Q. (2004) *Is Cyber-Harassment a Significant Problem? A Report On Children's Experience*, Calgary: University of Calgary.

——(2005) 'Cyber-harassment: A new method for an old behavior', *Journal of Educational Computing Research*, 32(3), 265–77, available at: http://www.ucalgary.ca/~qinli/publication/cyber-bully-Journal_of_Educational_Computing_Research_final.doc (accessed 27 April, 2011).

Campbell, M.A. and Gardner, S. (2005) 'Cyberbullying in high school', quoted in M. Campbell, 'Cyber-bullying – An old problem in a new guise', *Australasian Journal of Guidance and Counselling*, 15(1), 68–76, available at: http://eprints.qut.edu.au/archive/00001925/ (accessed 27 April 2011).

Grbich, S. and Wotherspoon, A. (2006) *National Safe Schools Week Report to DECS about Anti-Cyber Bullying Project*, South Australia.

Handshin, M. (2006) *2006 National Safe Schools Week Student Voice Roundtables State Summary and Student Feedback*, available at: http://www.deewr.gov.au/Schooling/NationalSafeSchools/Pages/nationalsafeschoolsframework.aspx.

Rigby, K. (2010) 'Evidence does not support the view that bullying is on the rise', *Australia Teacher Magazine*, Thursday 29 April, available at: http://www.deewr.gov.au/Schooling/NationalSafeSchools/Pages/nationalsafeschoolsframework.aspx (accessed 17 December 2010).

Chapter 17

A review of initiatives using technology to promote cyber-safety and digital citizenship

Barbara A. Spears

Introduction

Technology has rapidly become central to young people's socializing and communication strategies across the globe. Aligning with this progression, governments, non-government bodies, not-for-profit organizations and individuals world-wide have collaboratively and independently, expended increasingly significant time and resources on ensuring young people are adequately educated and prepared for life in a technologically wired society. Keeping young people safe online through cyber-safety education and supporting them to develop into ethical, digital citizens, where they can successfully navigate a rapidly changing technological environment, are two different sides of the same coin. This chapter outlines the cyber-safety initiatives undertaken in Australia over the past decade, as they have shifted from paper-based to CD-ROM to online, interactive websites in concert with the shifts in daily technology use. It considers digital citizenship as a construct, with a view to understanding how both of these approaches may contribute to positive uses of technology and safer relationships on and offline.

Overview

Cyber-safety is generally understood to mean safety in using or use of, the internet, or as Netsafe (www.netsafe.org.nz) suggests: 'the safe and responsible use of Information and Communication Technologies (ICTs)'. The Australian Parliament Interim Report of the Joint Select Committee on Cyber-Safety: *High-Wire Act: Cyber-Safety and the Young* (Parliament of Australia, 2011) reported that the term, though widely used, was 'largely undefined' and determined that it was 'a broad concept that concerns minimising the risks to children online from a range of negative influences including inappropriate social behaviours, abuse, identity theft and breaches of privacy' (ibid., p. 13). In short, cyber-safety refers to regulatory and educative approaches which are concerned with risk management and harm minimization strategies in relation to cyberbullying, cyberstalking, online grooming and sexting; privacy and identity theft. Digital citizenship, on the other hand, is a construct which embraces the ethics, values and behaviours required of citizens to operate appropriately in society, as applied to the online environment. It includes an understanding of the tools individuals need to operate effectively online, such as media literacy skills, as well as respecting the norms of behaviour regarding access and participation. Mossberger, Tolbert and McNeal (2008) referred to digital citizens as those who were online daily, and who had the educational competencies, means and the skills to participate. Digital citizenship, then, is concerned with the broader notion of what it means to operate responsibly and respectfully in the online environment, presumably, with an eye to one's own and others' safety online. Both of these notions support anti-cyberbullying and cyber-safe initiatives, as they aim to produce young people who can operate ethically, safely and effectively when online, through skills development, and the management of risk and minimization of harm.

In Australia, the National Safe Schools Framework (2011) places an emphasis on well-being and safety as essential elements supporting academic and social development, and the rights of all members of the school community to feel safe and be safe at school. In particular, it states: 'Australian schools actively support young people to develop understanding and skills to keep themselves and others safe' (Department of Education, Employment and Workplace Relations, 2011b, p. 5). Clearly, this is a proactive stance with regard to cyber-safety and digital citizenship initiatives, either through curricular strategies or through add-on interventions and programs and is highly relevant to contemporary educational settings whose prime role is to support effective learning.

The context

Australia is investing heavily in developing cyber-safe practices through the educative prevention, intervention and promotion initiatives from across government, non-government and not-for-profit organizations and has been active in this area for over 15 years. What is not known, however, is the individual, collective and cumulative efficacy of these to bring about change in attitudes or behaviours, as evaluations have either not been readily undertaken or built into cyber-safety initiatives until relatively recently. Or if they have been evaluated, have served to provide the organization with feedback, but may not have been published in a public forum. Given that the early initiatives had to operate in a rapidly changing public knowledge and awareness context, the question needs to be raised as to whether it matters if they had been evaluated at the time, as the needs of the community were in constant flux, marching from one new learning to another. From initially needing to learn about the role that the internet could play in their lives, to needing to develop protective strategies, to needing to understand the implications of social networking on young people's identity and relationship development, one initiative would be discarded almost as quickly as another was devised, resulting in a plethora of responses to perceived problems. Much of what has been done in the past has been stand-alone initiatives rather than a systematic, co-ordinated strategic approach which is more prevalent now. Evaluating this more systemic approach is indeed necessary, as it aims to change how society and school communities in particular, operate in, around and through technological developments. To contextualize these initiatives, however, the rapid technological change in the past two decades must be noted, as this has implications for what initiatives were developed when, their objectives, and the target audiences.

Historical snapshot

The first proposal for a World Wide Web was made in 1989–90 and the internet was eventually commercialized for public use in 1995. In 1989, internet access in Australia was only available to universities, and it was only this commercialization of the internet that enabled the public and individuals to have general access through Internet Service Providers (ISPs), initially via dial-up and then broadband services. In essence, then, 1995 saw the beginning of the need for cyber-safe initiatives in Australia, as the internet wove its way into business, schools and families, becoming more central to everything we did. Schools have had, and continue to have protective behaviour policies and strategies, which teach personal safety skills, and child abuse prevention strategies, with a view to education about violence, neglect and abuse. The shift to cyber-safe initiatives was a natural progression.

The historical development of this technological revolution from 1995 to 2011 is important, as each new technological development brought new challenges in terms of online behaviours and practices, requiring new awareness raising initiatives to educate the public of the latest technological iteration, and what it meant in terms of safety. Children born in 1995, are 16 years old in 2011: socializing and being educated in the middle of the mobile technology and social networking phenomena, having access to more information than ever before, moving seamlessly from offline to online environments, but seeing it as essentially the 'same' life (see Chapter 1 in this volume). Some teens will only ever know relationships that involve some online component. Parents of these young people do not have personal experiences of a teenage life like this, and so two key issues emerge: that we need to learn from young people about their experiences; and that we need to keep educating about the impact of technology on relationships in educational settings as technology changes, as they mature. Young people still need to grow and develop socially and emotionally through adolescence, and parents still need to parent them towards adulthood, and both need accurate information on how technology is influencing youth culture and their relationships with others.

The Alannah and Madeline Foundation in its submission to the Parliamentary inquiry on cyber-safety (2011, p. 50), noted that as at June 2010, 'more than 1.6 million young Australians had received cyber-safety lessons in the classroom … yet rates of cyberbullying, sexting, identity theft and breaches of privacy and sexual exploitation continue to rise'. This somewhat sobering statistic, suggests, on the one hand, that a lot has been done for a lot of young people, but while we may be educating and thus meeting the immediate needs of one cohort in time, there is a need for longitudinal studies, which track the ongoing and continued attitudes and behaviours of teens as they grow and mature with these technologies, which in turn provide new and different opportunities for misuse and abuse. In short, some milestones in the technological revolution need acknowledging, so that they contextualize this statement and the initiatives which have been undertaken to date in Australia.

The early internet, with its text-based search engines, static 'shopping cart' web pages and 'read only' capabilities, such as Netscape (1994); Hotmail (1996); Google (1998); Wikipedia (2001) and MySpace (arriving as a social networking site in 2003), represent Web 1.0 environments (Chapman, 2009). Cyber-safety initiatives implemented during this decade reflect the skills and knowledge awareness required to operate at this level. Everyone was learning: individuals needed to be educated about the capacity of the environment for positive and negative use, and be mindful of the information they gave out. Abuse and misuse occurred through phone, e-mail and sms texts. Mobile phones with cameras ready to connect to the internet were still being established; no YouTube was available to upload video footage to; no Facebook to de-friend others. From 2004, however, the term 'social media' became commonplace as did the notion of Web 2.0 environments, with their ever increasing interactive sharing and creating/publishing capabilities. Facebook emerged in 2004, soon overtaking MySpace as the social networking leader; and 2005 saw the launch of YouTube, a video/film sharing platform; 2006 brought Twitter and 2007 the iPhone and mobile web, Tumblr, Kindles and Google Street View; the beginnings of Cloud technology emerged in 2008; Google Docs appeared in 2009, Formsping, an anonymous question and answer-based social site and cloud gaming platforms arrived in 2010. Google+ is currently being tested and Photovine, Google's photograph-based social networking site was released in July 2011.

Our current 16-year-olds, born in 1995, when the internet was first commercialized for public use, have grown up totally surrounded by and engaged with this rapid change and are now peak users. Parents have provided the access to this technology, but have perhaps not been so quick to comprehend its capabilities and social impact. Clearly cyber-safety initiatives in this user/creator environment still require information and awareness raising, but they also require additional strategies which build on previous work and understandings of adolescent and group development. Adopting an integrated, developmental framework that sees safety as a key focus for younger children, merging into online ethical behaviours and effective decision-making as they grow and develop, recognizes that younger children have different safety needs to those who are older. Addressing peer relationships and friendship-driven practices (Ito *et al.*, 2008) in networked publics (boyd, 2007) need consideration, as youth culture has changed in concert with technological change. Young people are 'hanging out' in online public domains, the 'malls' of cyberspace. In Web 2.0 and Web 3.0, where one's profile online is fluid and mobile and the user is also the publisher, comes the sober realization that *what goes online, stays online*. With this in mind, government and non-government cyber-safety initiatives are outlined.

Australian government cyber-safety initiatives

In 1999, *NetAlert Ltd*, Australia's Internet Safety Advisory body (http://www.ozguide.com/netalert_teachers_guide.pdf), was established with the clear objectives of: creating a helpline for parents and families; creating an information resource for internet users, including support for research; collaborating closely with industry, communities and regulatory bodies; and initiating a program of community education (Parliament of Australia, 2000–2001). These simple initiatives are clearly about providing information and access to the public. Following an audit of internet safety by the Curriculum Corporation, however, it was determined that there were few resources for schools and that there was 'little that was consistent in the way that internet safety was taught'. As a result, in 2004, *CyberSafe Schools* was launched (Curriculum Corporation, 2005), a three-phase initiative, which saw resources delivered to every school in Australia: comprising:

- a wall chart, a Quick Reference Guide and a Teacher's Guide to Internet Safety (http://www.ozguide.com/netalert_teachers_guide.pdf);
- a primary school-focused CD-ROM called *Cyberquoll* featuring animations and interactive games addressing netiquette; (http://www.cybersmart.gov.au/cyberquoll/);
- a secondary school interactive program: *CyberNetrix* (http://www.cybersmart.gov.au/cybernetrix/index.html).

As well as student activity sheets, and teacher guides, a teacher survey was included to assess the program and gather feedback. Of interest, is that both *Cyberquoll* and *CyberNetrix* have survived and been incorporated into the current online education strategy (see cybersmart.gov.au).

What follows is a brief look at the fast moving progression of mergers, updates and initiatives since that first entrée into cyber-safe education and awareness-raising, demonstrating that the Australian government was endeavouring to keep up with the speed of the challenges presented by new and emerging technologies.

In 2006, *Cybersmart Kids* an online community awareness project was launched by the Australian Communications and Media Authority (ACMA) (Australian Communications and Media Authority, 2006, Annual Report) and early in 2007, *NetAlert* merged with ACMA as part of the *Protecting Australian Families Online* initiative, which aimed to protect Australian families, through strategies such as: filtering, law enforcement, increased regulation, support, education and awareness initiatives. Suddenly, protective behaviours were no longer only about violence, child abuse and neglect in the offline environment, but were an important strategy in the online armoury. According to the *High-Wire Act Cyber-Safety Report of the Joint Select Committee on Cyber Safety* (Australian Communications and Media Authority, 2011, p. 15), the Australian Communications and Media Authority has operated in the 'cyber-safety space' for over 10 years, in its role as the regulatory body for broadcasting, internet, radio and telecommunications, investigating complaints about prohibited online content.

Between 2007 and 2008, ACMA updated *Cybersmart Kids* to include information on safe social networking and cyberbullying. *WiseUpToIt* (http://www.cybersmart.gov.au/wiseuptoit/), a resource which used videoclips of four young people telling their stories, along with downloadable scripts, lesson plans and student handouts was added as a secondary resource (Australian Communications and Media Authority, 2007, Annual Report). Clearly, between 1999 and 2008, several significant, but effectively stand-alone initiatives were undertaken through these key government departments in an effort to raise awareness and educate the public about online risk. Little published evaluation of these is available, however.

With the announcement of the National Cyber-Safety Plan by the Department of Broadband, Communication and Digital Economy (DBCDE) in 2008, $AUD 125.8 million over four years was committed to developing a comprehensive approach to combat online risk and support parents and educators through awareness-raising and positive initiatives. Included was funding for the establishment of a Consultative Working Group (2008) to advise government on 'protecting children from online risks from cyberbullying, exposure to illegal content and privacy breaches'; and the formation of a Youth Advisory Group (2009) to provide young people's perspectives and ongoing research into the changing digital environments to identify issues. In 2011, this group was scheduled to include over 1200 young people from 120 schools, a considerable increase from the 300 initially involved in 2009. This recent strategy, unlike previous stand-alone resource development, was multi-faceted in approach, was employing youth voice to advise on strategies and policy, and was working with an educative strategy through ACMA. A new web presence, to take the place of *NetAlert* and *Cybersmart Kids* was developed, along with school internet safety presentations and professional development sessions for teachers for a cyber-safety outreach program. Launched in July 2009, the cybersmart (http://www.cybersmart.gov.au/) website provides a 'one-stop shop' for cyber-safety information and resources for children of varying age groups, parents and teachers and a schools' portal was also developed. The *Cybersmart online helpline* was launched in 2010, offering real-time counselling and advice to young people and the *Cybersafety Outreach* program, incorporating professional development for teachers and internet information presentations to schools and families expanded. *Hector's World* partnered with ACMA to provide online safety resources for young children, and *SuperClubsPLUS* Australia, a protected social learning environment for children aged 6–12 years (http://www.scplus.com/d/index.php), has also partnered with ACMA to develop and deliver a 'Cybersmart Badge'. In addition, *Cybersmart Detectives* (http://www.cybersmart.gov.au/Schools/Teacher%20resources/Upper%20primary/Interactive%20Shared%20 Learning/Cybersmart%20Detectives.aspx), an interactive online activity was played by over 6,500 students during 2008–09. The Outreach program was extended to include pre-service teachers at university and in 2010, nearly 123,000 teachers, parents and students had participated. An independent review of this program was scheduled for 2010–11 and also for the *Cybersmart Detectives* program, recognizing the importance of establishing efficacy of the message reach and delivery to the target groups. *Let's Fight It Together*, adapted from the UK ChildNet organisation was also launched in 2009, providing customized advice, lesson plans and interviews with Australian children. In October 2010, an interactive cyber-safety guide for parents was released and in May 2011, *Connect-ed*, a free, online training module for teachers was launched, covering such areas as: cyberbullying, sexting, excessive internet use, mobile use, e-security and managing digital reputations. *Cybersmart Access*, for children with special educational needs; *Tagged*, a film resource and *Cybersmart Networking* are the most recent additions to the portfolio (November, 2011).

From 2007, when *NetAlert* merged with ACMA, until now, there has been a flurry of activity concerning cyber-safety, mainly through ACMA's cybersmart initiative, the development of which has been underpinned by an evidence base provided by research activities since 2000 into: parent concerns; young people's use of social media; digital media literacy; media and communications in families; and developments in internet filtering (http://www.cybersmart.gov.au/Schools/Research/ACMA%20Research.aspx).

In 2009, the Department of Broadband, Communications and the Digital Economy (DBCDE) specifically commissioned a literature review of existing Australian and international cyber-safety research, which found that there were 'significant gaps in most areas [of cyber-safety research]' (Dooley *et al.*, 2009, p. 9) particularly in Australia, and that few of the resources available up to that time, had been 'empirically validated' (ibid., p. 187).

In light of this lack of evaluation of resources and innovations, the DBCDE has commenced a series of surveys to determine 'the changes in awareness and behaviour in relation to cyber-safety and relevant e-security risks': effectively, to determine the efficacy of the intervention approaches to date. The subsequent *Australian Children's Cyber Safety and e-Security Report* (IRIS Research and the Australian Council for Educational Research, 2010, ongoing) into the nature and prevalence of risks facing children using the internet, has the express aim of informing policy responses and the ongoing delivery of effective communication strategies. Two reports have been completed: parents' perceptions and knowledge of their children's experiences using the internet (June 2010) and teachers' experiences and knowledge (March 2011). Specific objectives related to how parents responded to cyber-safety, including: 'awareness and use of government cyber-safety and e-security initiatives' (ibid., p. 7) and assessing teachers' 'level of awareness and usage of existing cyber-safety resources; and to identify the existing resources that are considered to be the most helpful' (ibid., p. 8). This research provides baseline data and will enable the ongoing tracking of change in levels of awareness and behaviours, providing the first evidence of the success or otherwise of various cyber-safety initiatives within the cyber-safety plan, which have been implemented since 2008.

Findings from the Iris Reports (IRIS Research and the Australian Council for Educational Research, 2010, 2011) from a representative, national, random sample of parents (N = 2014) and teachers (N = 2088) indicated that the majority of parents had spoken to their children about the risks of being online (82.4 per cent) and 80.3 per cent had implemented preventative measures to reduce risks to their children. Almost 1 in 2 (46.1 per cent) felt well informed about cyber-safety issues, with a further 33.2 per cent feeling 'somewhat informed' (IRIS Research and the Australian Council for Educational Research, 2010, p. 2). The teachers' survey found that 9 out of 10 teachers (90.1 per cent) applied cyber-safety practices to their personal use of the internet (IRIS Research and the Australian Council for Educational Research, 2011, p. ii), with over one-third (36.1 per cent) feeling 'well informed' and 51.9 per cent feeling 'somewhat informed'. The most common forms of cyber-safety incidents in schools in the previous 12 months were: students receiving threatening or bullying e-mails, telephone calls or SMS messages (39.6 per cent); accessing inappropriate websites or material (25 per cent) or personal/private material posted on social networking sites (24.6 per cent) (ibid., p. iii). In the past 12 months, 59.3 per cent of teachers had accessed some sort of information about cyber-safety, with 71.9 per cent 'feeling satisfied' and 21.5 per cent feeling 'highly satisfied' with the information. The most highly recognized information sources were the *Cybersmart* website; the *Bullying. No Way!* website and their local Education Department training information. The *Cybersmart* website was considered the most helpful (41.3 per cent) followed by the local

Department of Education information (23 per cent) and the *Bullying. No Way*! website (17 per cent). These studies go part way towards addressing the gaps in Australian cyber-safety research, as previously noted by Dooley *et al.* (2009) and highlight that cyber-safety messages are being heard by parents and teachers.

The Senate Inquiry into cyber-safety (Parliament of the Commonwealth of Australia, 2011) undertook two online surveys of young people in relation to cyber-safety (*Are You Safe?*). A total of 33,751 young people completed surveys, with 18,159 aged 12 years and under; and 15,592, 13–18-year-olds. In terms of young people's responses to where they have learnt about cyber-safety (ibid., p. 277), 41.6 per cent reported that they learnt about it at school; 30.7 per cent from family; 13.6 per cent from friends; information on the internet (8.3 per cent); other (3.9 per cent) and surprisingly, 2.0 per cent said they had never learnt anything about it. Further to this, 49.6 per cent reported that they had increased their privacy settings of their social networking sites to the highest level, but 20.2 per cent reported having left them at the default settings, with some 9.9 per cent not having explored them at all, and a further 4 per cent had deliberately disabled their privacy settings to allow everyone access (ibid., p.155). The rest either did not use social networking sites (11.6 per cent) or did not know (4.7 per cent).

Clearly, young people are learning about cyber-safety from a variety of sources, but mostly from school and/or home, with very few accessing the internet for information. That many are taking responsibility for protecting their online profile is indicative that the messages about personal safety may be getting through. For some though, leaving trust in the provider to have settings set at an adequate level seems naïve. Asked if they shared their name online, however, the overall majority did, with older students (75 per cent of 17-year-olds) more likely than younger students (55 per cent of 12-year-olds) to do so. Given that Facebook uses names to find friends, this is not surprising, but is indicative of a perception that perceived risk may be relative to the audience: i.e. friends on Facebook do not present risk. A different story, for example, was evident for their address, and telephone number, with 93.2 per cent reporting that they would not divulge their address; or phone number (90.4 per cent) online. Longitudinal research such as that being started by DBCDE is needed on the messages that young people are hearing about cyber-safety, but this author suggests that schools may no longer be the prime sites for determining reach and distribution of the cyber-safety messages. Accessing young people where they socialize and play online, through online social marketing campaigns needs consideration (Spears and Zeederberg, in press). Clearly they are not actively seeking out cyber-safety information on the internet, as only 8.3 per cent indicated that they did this, which raises the question of whom these websites are actually serving, and if the viewing hits recorded, are really the target audience of young people. Knowing that they are constantly visiting such sites as

YouTube, and online gaming sites, social network sites and other networked public domains, means that getting the message directly to them, via online social marketing techniques is an option. Being able to track users through web analytics can clearly establish reach and impact.

Other government departments, however, also have some regulatory and policy responsibilities in relation to cyber-safety. Through DEST, the forerunner to the Department of Education Employment and Workplace Relations (DEEWR), two major research projects were commissioned under the National Safe Schools umbrella, to explore the prevalence and nature of covert and cyberbullying amongst Australian students (Cross *et al.*, 2009; and Spears *et al.*, 2008; 2009). These established the first studies into covert bullying, with one providing a cross-sectional study of the phenomenon across Australia, and the second providing a qualitative, information rich exploration of the human dimension and impact of it through youth stories of covert and cyberbullying.

The revised National Safe Schools Framework (2011) was launched to coincide with the National Day of Action Against Bullying and Violence, and has a clear emphasis on the role that the online environment now plays. Most recently, through the *Digital Education Revolution* (Department of Education, Employment and Workplace Relations, 2011a) $2.4 billion is being provided over 7 years, to prepare students for future education, training and to live and work in a digital world and has provided access to computers for all Year 9–12 students; is deploying high speed internet connections to schools; is collaborating with State/territories to ensure all new and continuing teachers have access to training in the use of Information and Communication Technologies (ICTs); has maintained the *Bullying. No Way!* website (http://www.bullyingnoway.com.au/); has provided access to a *Cyber-Safety Help Button* (http://www.dbcde.gov.au/online_safety_ and_security/cybersafetyhelpbutton_download) and has funded the *eSmart* program, a national pilot to increase cyber-safety in schools (http://www.deewr.gov.au/Schooling/DigitalEducationRevolution/Pages/AMFoundation). *eSmart* assists schools in adopting cultural norms of smart, safe and responsible digital technologies through focusing on the positive ways to use technology in school environments.

Other government departments such as the Australian Federal Police; the Attorney-General's Department; and the Australian Institute of Criminology all engage with various aspects of cyber-safety issues, such as: cyber-racism; identity security; grooming and procuring offences; law enforcement and child protection matters; and technology-enabled crime, but discussion of these programs are beyond the parameters of this chapter. In addition, Federal initiatives such as *KidsMatter* (http://www.kidsmatter.edu.au/) and *Mindmatters* (http://www.mindmatters.edu.au/default.asp), which support the social and emotional development of young people in schools to improve mental health and well-being are important, well-evaluated components of the whole community approach required to develop social capacity overall.

State government and territory cyber-safety initiatives

At the State and Territory (Northern Territory and the Australian Capital Territory) level, a range of policies, initiatives and programs are available to support young people being safe online, which are mainly offered through the various Departments of Education, Independent Schools and Catholic Education sectors and represent a multi-faceted strategy covering: peer relations, anti-bullying, public awareness initiatives for parents and caregivers, and access for vulnerable children and youth (see ACMA http://192.43.227.232/Schools/Cybersafety%20policies%20and%20procedures/National%20or%20state%20and%20territory%20policies.aspx).

In brief, all States have representatives on the *Safe and Supportive School Communities* project, a national collaboration which is linked to the *Bullying. No Way!* website, which provides a wide range of anti-bullying materials. Several States have senior level committees which oversee initiatives such as the Coalition to Decrease Bullying, Harassment and Violence in South Australian Schools (2005); the Queensland Schools Alliance Against Violence (2010); and the Western Australian Cyber Safety for Children Working Party. Some states support Peer Mediation programs. Most have acceptable use policies, where students and parents acknowledge that they have a responsibility to use the internet and other services safely, and are aware that if they breach the guidelines, then disciplinary action or restricted access may occur. Anti-bullying policies exist as either a stand-alone document, or included as part of their overarching behaviour management documents. Many, through the *Digital Education Revolution*, have access to secure wireless networks and laptops in schools, which are subject to strict internet filters. For a comprehensive review of what is currently being done in each State in terms of cyber-safety initiatives, please consult the Joint Select Senate Committee Interim report on cybersafety: *High-Wire Act: Cyber-Safety and the Young* (Parliament of the Commonwealth of Australia, 2011). What is evident, is that through each State's education department policies, initiatives and programs, and in conjunction with the Federal government initiatives previously outlined, Australian children and youth are well provided for through a multi-layered mechanism of strategies and approaches involving weblinks and resources. What is not known, is how effective these strategies are nor whether they might be contributing to the notion of digital citizenship.

Non-government initiatives

There are many not-for-profit organizations and individuals who have also entered the cyber-safety space, making it very crowded indeed. There are some who advocate against mandatory internet filtering as a strategy, such as the *Safer Internet Group* and some who actively aim to help young people live happier, mentally healthy lives, such as the *Inspire Foundation* (http://www.inspire.org.au/). Through their *Technology and Wellbeing Roundtable*, funded by the Telstra Foundation, they actively bring together leaders from industry and research who collaborate to share insight into how technology can enable young people's lives. There are those such as *Berry Street*, an independent family welfare organization who offer services such as *BeNetWise*, an approach to cyber-safety and digital inclusion to the vulnerable and socially isolated (http://www.berrystreet.org.au/BeNetWise). The *Alannah and Madeline Foundation*, a national children's charity aimed at keeping children safe from violence, operates the *National Centre Against Bullying*, and the *eSmart* framework for cyber-safety and well-being in schools.

Universities continue to play a significant role in providing research and resource development which others can use through the development of the Australian Universities' Anti-Bullying Research Alliance (AUARA). The Edith Cowan University in Western Australia has developed an evidence-based cyberbullying prevention resource, *Friendly Schools*, due for release in 2011. Other universities are engaged in either the production of evidence-based resources, e.g. Flinders University's DVD on *Coping With Bullying;* or through research initiatives aimed at improving schools' approaches to dealing with bullying and cyberbullying, e.g. The University of South Australia's evaluation of the *Method of Shared Concern*; a study to explore young people's cyber-bystander behaviour; and the use of online social marketing as an educative and intervention strategy. Australian Research Council collaborations, such as that led by the Queensland University of Technology into legal issues surrounding cyberbullying and possible interventions to equip young people to understand the law in relation to cyberbullying are also being undertaken.

Of significance, the not-for-profit Inspire Foundation successfully led a bid to establish a Cooperative Research Centre for Young People, Technology and Wellbeing (YAW-CRC) which came into action in July 2011. Funded by the Department of Innovation, Industry Science & Research (DIISR) with $AUD 27 million and supported by over 7 partner organizations, this initiative will promote cyber-safety, mental health and well-being over the next five years through three programs: (1) *Safe and Supportive*, which uses technologies as settings to promote cyber-safety and strengthen resilience, mental health and well-being of all young people; (2) *Connected and Creative* which aims to use technologies to help those vulnerable and marginalized learn the skills required and have access; and (3) *User Driven and Empowered* focusing on how technologies can act as facilitators of good mental health for those experiencing mental health problems. For more details, please visit http://www.yawcrc.org.au/wp-content/uploads/2011/07/CRC_Overview_9July2011.pdf. This initiative brings together an end-user driven, youth-oriented, multi-faceted, cross-sectoral and educative approach with a clear focus on using technology positively to deliver cyber-safety and well-being messages for young people, by young people and with young people and is an example of the collaborative approach needed to effect social change.

The *National Association for the Prevention of Child Abuse and Neglect (NAPCAN)* offers a range of educative online programs and campaigns. The *SOSO (Smart Online, Safe Offline* http://www.soso.org.au/) offers a stranger danger and a cyberbullying component which targets bystanders. Of particular interest is the collaboration with digital marketing group *Zuni* to produce an online social marketing campaign to reduce cyberbullying, where young people were challenged in their online social networks, YouTube and gaming sites to take a stand against cyberbulling using the campaign phrase: *Cyberbullying Affects Real Lives. SOSO* is now being incorporated in the above Collaborative Research Centre's *Safe and Supportive* program as a five-year longitudinal study into online social marketing and this author suggests that this approach could be the way of the future for some cyber-safety initiatives: viral marketing campaigns, online databases and sites where traffic, reach and impact can be measured and tracked through web analytics (Spears and Zeederberg, in press). Taking the cybersafety message to young people as they live, work and play online, in their social networking and gaming sites and using current technologies to do so, rather than separating it to offline school-based components and online websites that they may rarely visit, would seem sensible, and indeed acknowledges what young people are telling us: that they are not separating their on and offline lives as we do, but are seamlessly moving to and fro between them as if it were the same life.

Digital citizenship

The construct known as *Digital citizenship* is a term growing in acceptance but has not been widely addressed in Australia to date in relation to specific programs and initiatives. However the New South Wales Department of Education and Training (DET) has established an online, interactive digital citizenship education program, designed to teach students the skills required to be good digital citizens. Referring to work done by *digizen.org* (UK) and *digitalcitizenship.net* (USA) and bringing together resources from *Cybersmart* and others such as *budd-e* from the Australian cyber-security website (https://budd-e.staysmartonline.gov.au/secondary/main.php), and *ThinkUKnow* from the Australian Federal Police (http://www.thinkuknow.org.au/), DET is actively targeting cyberbullying and promoting the expectation that as good digital citizens, all students should be active in preventing it. Digital citizenship, can thus be construed as an overarching construct, embracing all aspects of cyber safe practices required to produce skilled individuals who can participate fully online.

The Melbourne Declaration of Educational Goals for Young Australians (Ministerial Council Education, Employment and Youth Affairs, 2008) commits 'to supporting all young Australians to become successful learners, confident and creative individuals, and active and informed citizens'. Further to this, the Australian Curriculum, Assessment and Reporting Authority (http://www.acara.edu.au/curriculum/curriculum.html) notes that:

- successful learners: are creative and productive users of technology;
- confident learners: develop personal values and attributes such as honesty, resilience, empathy and respect for others and have the knowledge, skills, understandings and values to establish and maintain healthy and satisfying lives and
- active and informed citizens: act with moral and ethical integrity.

Having the skills to particiate is only part of the cyber-safety solution. Having a moral compass and personal values which sit alongside these skills, in relation to online behaviours, relationships with others in public networks and enabling full participation in a technologically driven world, would seem to be of equal import. The cyber-safety initiatives noted in this chapter predominantly deal with providing information designed to raise awareness and educate, and shift attitudes and behaviours in concert with ever changing ICTs. What is now needed, are successful, confident young people who are active and informed *citizens*, who act with moral and ethical integrity, capable of using technology for positive rather than negative outcomes. If their online and offline worlds really are the same space, then young people not only need to be skilled in access, communication, and security, but also in respect, inclusion and etiquette. Digital citizenship may reflect the mature phase of our endeavours, that which only occurs when young people have learnt how to operate safely and effectively, enabling them to take responsibility for themselves and to care about others.

References

Alannah and Madeline Foundation (2011) *National Centre Against Bullying*, available at: http://www.amf.org.au/NCAB/ (accessed May 2011).

Australian Communications and Media Authority (2006) *Annual Report* (*2006–2007*) available at: http://www.acma.gov.au/WEB/STANDARD/pc=PC_310778 (accessed May 2011).

——(2007) *Annual Report (2007–2008)*, available at: http://www.acma.gov.au/WEB/STANDARD/pc=PC_311428 (accessed May 2011).

——(2009a) *Click & Connect: Young Australian's Use of Online Social Media*, available at: http://www.acma.gov.au/webwr/aba/about/recruitment/click_and_connect-02_quantitative_report.pdf (accessed May 2011).

——(2009b) *Annual Report (2009–2010)*, available at: http://www.acma.gov.au/WEB/STANDARD/pc=PC_312295 (accessed May 2011)

——(2011) *Publications and Research Programs*, available at: http://www.cybersmart.gov.au/Schools/Research/ACMA%20Research.aspx (accessed April 2011)

boyd, d. (2007) *Why Youth (Heart) Social Network Sites: The Role of Networked Publics in Teenage Social Life*, Cambridge, MA: MIT Press.

Chapman, C. (2009) *History of the Internet in a Nutshell*, available at: http://sixrevisions.com/resources/the-history-of-the-internet-in-a-nutshell/ (accessed May 2011).

Cross, D., Shaw, T., Hearn, L., Epstein, M., Monks, H., Lester, L., and Thomas, L. (2009) *Australian Covert Bullying Prevalence Study* (ACBPS), Perth: Child Health Promotion Research Centre, Edith Cowan University.

Curriculum Corporation (2005) 'Cyber Safe schools', *Curriculum Leadership* (3), Issue 4, available at: http://www.curriculum.edu.au/leader/cybersafe_schools,8063.html?issueID=9737 (accessed May 2011).

Department of Education, Employment and Workplace Relations (2011a) *Digital Education Revolution: Cybersafety in Schools*, available at: http://www.deewr.gov.au/Schooling/DigitalEducationRevolution/Pages/Cybersafetyinschools.aspx (accessed April, 2011)

——(2011b) *National Safe Schools Framework*, available at: http://www.deewr.gov.au/Schooling/NationalSafeSchools/Pages/nationalsafeschoolsframework.aspx#vision (accessed May 2011).

Department of Education and Training (2010) *Digital Citizenship*, available at: http://www.digitalcitizenship.nsw.edu.au/ (accessed April 2011).

Dooley, J.J., Cross.,D., Hearn, L. and Treyvaud, R. (2009) *Review of Existing Australian and international cyber-safety research*. Perth: Child Health Promotion Centre, Edith Cowan University.

IRIS Research and the Australian Council for Educational Research (2010) *Australian Children's Cybersafety and E-Security Report: Parents' Survey for the Department of Broadband, Communications and the Digital Economy*, available at: http://www.dbcde.gov.au/–data/assets/pdf_file/0004/132691/Security-Projectreport_on-the-results-of-a_parents_survey.pdf (accessed May 2011).

——(2011) *Australian Children's Cybersafety and E-Security Report; Teacher's Survey for the Department of Broadband, Communications and the Digital Economy*, available at: http://www.dbcde.gov.au/ – data/assets/pdf_file/0003/135561/Australian_Childrens_Cyber-safety_and_E-Security_Projectreport_on_the_results_of_a_teachers_survey.pdf (accessed April 2011).

Ito, M., Horst, H.A.S., Bittanti, M., boyd, d., Herr-Stephenson, B., Lange, P.G., Pascoe, C.J. *et al.* (2008) *Living and Learning with New Media: Summary of Findings from the Digital Youth Project* (the John D. and Catherine T. MacArthur Foundation Report on Digital Media and Learning), available at: http://mitpress.mit.edu/catalog/item/default.asp?ttype=2&tid=11940 (accessed July 2011).

Ministerial Council Education, Employment and Youth Affairs (2008) *Melbourne Declaration of Educational Goals for Young Australians*, available at: http://www.curriculum.edu.au/verve/_resources/National_Declaration_on_the_Educational_Goals_for_Young_Australians.pdf (accessed April 2011).

Mossberger, K., Tolbert, C.J. and McNeal, R.S. (2008) *Digital Citizenship: The Internet, Society and Participation*, Cambridge, MA: The MIT Press.

NetAlert (2004) *Cyber Safe Schools: A Teacher's Guide to Internet Safety* available at: http://www.ozguide.com/netalert_teachers_guide.pdf (accessed May 2011).

Parliament of the Commonwealth of Australia (2000) *Senate Environment, Communications, Information Technology & The Arts Legislation Committee Estimates Questions*, available at: http://www.aph.gov.au/senate/committee/ecita_ctte/estimates/add_0001/cita/05bip.pdf (accessed May 2011).

——(2011) *High Wire Act: Cyber Safety and the Young*. Joint Select Committee on Cyber Safety, available at: http://www.aph.gov.au/house/committee/jscc/report.htm (accessed July 2011).

Spears, B.A. (2011) 'Developing innovative interventions for adolescents: methodologies for accessing youth voice', paper presented at the 2nd Research Training School: COST Meeting, Turku, Finland, 5 May.

Spears, B.A. and Zeederberg, M. (in press) 'Developing innovative intervention strategies for cyberbullying: Youth voice and online social marketing'. In S. Bauman (ed.) *Principles of Cyberbullying Research: Definition, Measures and Method.* London: Routledge.
Spears, B.A., Slee, P.T., Owens, L. and Johnson, B. (2008) *Behind the Scenes: Insights into the Human Dimension of Covert Bullying*, Report prepared for DEEWR: Canberra.
——(2009) 'Behind the scenes and screens: insights into the human dimension of covert and cyberbullying', *Zeitschrift für Psychologie* [Journal of Psychology], 217: 189–96.

Conclusion

Barbara A. Spears and Angela Costabile

The main aim of this book has been to provide an opportunity for the positive impact of technology on relationships in educational settings to be explored and celebrated. In doing so, this book provides a counterpoint to publications which have examined and highlighted the negative impact of ICTs on young people, for example, those focused on cyberbullying and cybersafety.

Cyberbullying is an online, social relationship problem, which requires relationship solutions if it is to be reduced and managed successfully. Therefore, while educating students, teachers and parents on how to manage technologies and behaviours associated with those technologies in educational environments will undoubtedly have some impact on cyberbullying, removing excessive filtering, controlling and abstaining from technology will not resolve cyberbullying problems on its own.

The technology which enables cyberbullying's very existence, is ubiquitous, rapidly converging and changing, and the world will never be as it was before, because of this technological and social change. Hence, embracing and understanding how we can harness the technologies to assist and support relationships in educational settings are paramount. Education and interventions which employ technology as part of the solution must be the way forward. One step in this process is to recognize the (positive and negative) ways in which technologies impact on young people and their relationships, and how we can support them now.

Static, read-only Web 1.0 platforms, which mastered the technical challenge of being online, gave way to interactive Web 2.0 platforms, which saw users become creators and publishers of content and online networks become central to young people's lives. Web 2.0, which brought social change through the advent of social networking sites and interconnectivity, is now heralding the emergence of Web 3.0, which some are referring to as the *personal* web, or the *semantic* web.

This book has captured what is happening as young people and educators have gone online during the time of Web 2.0, and will form a substantial basis of comparison for the transition to the next iteration: Web 3.0. It has brought 17 chapters together to focus on the impact of technology on relationships in educational settings, highlighting the positive aspects and to advocate for social supports and learning opportunities which can be delivered through technology. Authors have indicated that this can be done directly: for example, in classrooms with mobile technologies to improve learning as well as with the use of online games to change behaviours and attitudes. It can also occur indirectly: through the training of adults (teachers and parents) or children.

We have read how young people are interpreting their online worlds, how the curriculum and pedagogy are changing, and how identity and security need to be considered. We have seen evidence from educational settings of how learning is being enhanced through the use of mobile technologies, and how technology is being deployed in ways that develop the skills needed to establish and maintain peer relationships in the twenty-first century, as well as how it can support those whose relationships are failing. Finally, we have seen how evidence-based interventions can contribute to positive relationship outcomes, in the quest to curtail bullying and cyberbullying behaviors.

Young people have often been referred to as *digital natives* (Prensky, 2001), but does this necessarily make them skilled in handling relationships that are enacted online? More recently, the notion of *digital wisdom* (Prensky, 2009) has been presented, suggesting that the former *natives*, now have experiences from operating with ICTS, from which we so-called '*immigrants*' can learn. It does not necessarily follow, however, that they are more capable of managing their relationships, particularly as these occur in and across educational settings. Adolescents and young people are navigating these new mediated public arenas, and while they may be skilled in executing associated technologies, they are still teenagers, whose brains are still developing, whose execution of decision-making is likely to be flawed, whose peer networks exercise influence and control and who therefore need adult guidance and intervention.

It is important therefore to consider some key points in relation to this technological revolution that have been occurring during the transition from Web 1.0 to Web 2.0 to Web 3.0 and beyond. Young people may well be active learners, problem solvers, highly sociable, mobile and creative, and they need to be to operate in contemporary educational and social domains where ICTs are central and ubiquitous. But what of the impact of ICTs on socio-relational and cognitive-emotional experiences from a very young age? How has technology impacted on and changed the developing brain? What is the changing role of school and teachers in a world that is wired and where information can be accessed immediately? How ready are schools to manage the always-on relationships of their students? What does this mean for pre-service teacher education programmes? How has technology contributed to pedagogies and curricula design? What of the impact of Web 3.0 on relationships in the future?

In the past few years, different and opposing theories about the relationship between ICTs and young people have been suggested (Eco, 2001):

- the 'Apocalyptic' point of view, that blames ICTs as petitioners of educational failures;
- the 'Integrated' perspective, which considers mass media as holders of democracy due to its capability of spreading knowledge and information to everyone.

Through contributions from youth voice, young people have identified that online and offline environments are the *same* social space for them, through which their friendship-driven practices (Ito *et al.*, 2008) operate. This has implications for how relationships operate and are managed in educational settings. It also has implications for future research. Not to pay respect to youth voice, when they are the experts in their own online lives, risks alienating the very sources that are needed to understand how ITCs occupy the centrality that it evidently does for young people.

ICTs, then, are rewriting how we interact socially, forcing reconsiderations and reassessments of the social boundaries and mores within which we all operate, and of the relationships which operate within and around them. ICTs are enablers of young people: they provide resources, and opportunities to learn, and to make new things possible. Future researchers will reflect on the uniqueness of the situation in which society finds itself today. The technological revolution will reverberate for a while to come, as it shifts yet again through technological developments, but marking that period in time, through an international research lens such as that provided through this book, when young people were first learning to live, laugh, work and play in both online and offline environments, is a significant foundation from which to grow greater understanding of the positive impact that technology can have on relationships in educational settings.

References

Eco, U. (2001) *Apocalittici e Integrati*, Milan: Bompiani.

Ito, M., Horst, H.A.S., Bittanti, M., boyd, d., Herr, B., Lange, P.G., Pascoe, C.J. *et al.* (2008) *Living and Learning with New Media: Summary of Findings from the Digital Youth Project* (the John D. and Catherine T. MacArthur Foundation Report on Digital Media and Learning), available at: http://mitpress.mit.edu/catalog/item/default.asp?ttype=2andtid=11940 (accessed July 2011).

Prensky, M. (2001) 'Digital natives, digital immigrants', Parts I and II, *ON the Horizon*, 9(5).

——(2009) 'H. sapiens digital: from digital immigrants and digital natives to digital wisdom', available at: http://www.innovateonline.info/pdf/vol5_issue3/H._Sapiens_Digital–from_Digital_Immigrants_and_Digital_Natives_to_Digital_Wisdom.pdf (accessed 15 December 2010).

Index